Feminist Geograph

For the feminist geographers – students, colleagues, and friends – whom I've worked with and learned from over the years.

Feminist Geography in Practice: Research and Methods

Edited by Pamela Moss

BLACKWELL *Publishers*

Copyright © Blackwell Publishers Ltd 2002
Editorial matter and arrangement copyright © Pamela Moss 2002

The moral right of Pamela Moss to be identified as author of the editorial material has
been asserted in accordance with the Copyright Designs and Patents Act 1988.

First published 2002

2 4 6 8 10 9 7 5 3 1

Blackwell Publishers Ltd
108 Cowley Road
Oxford OX4 1JF
UK

Blackwell Publishers Inc.
350 Main Street
Malden, Massachusetts 02148
USA

All rights reserved. Except for the quotation of short passages for the purposes of
criticism and review, no part of this publication may be reproduced, stored in a
retrieval system, or transmitted, in any form or by any means, electronic, mechanical,
photocopying or otherwise, without the prior permission of the publisher.

Except in the United States of America, this book is sold subject to the condition that it
shall not, by way of trade or otherwise, be lent, resold, hired out, or otherwise
circulated without the publisher's prior consent in any form of binding or cover other
than that in which it is published and without a similar condition including this
condition being imposed on the subsequent purchaser.

British Library Cataloguing in Publication Data
A CIP catalogue record for this book is available from the British Library.

Library of Congress Cataloging-in-Publication Data has been applied for

ISBN 0-631-22019-4 (hbk)
ISBN 0-631-22020-8 (pbk)

Typeset in 10 on 12pt Sabon
by SetSystems Ltd, Saffron Walden, Essex
Printed in Great Britain by MPG Books Ltd, Bodmin, Cornwall

This book is printed on acid-free paper.

Contents

Acknowledgements

This project is the result of much effort by many people. I first want to thank all the authors for being sensitive to the goals of the project while writing their pieces and then taking my comments in the spirit in which they were given. I appreciate your patience and your writing skills! I enjoyed tremendously being part of the Feminist Pedagogy Working Group. The vim and tenacity of all these women provided me with endless hours of inspiration and learning. I thank them all individually for their work and effort – Amy Zidulka, Andrea Lloyd, Denise Pritchard, Jenny Kerber, Kathleen Gabelmann, Cristal Scheer, Erin Quigley, Joy Beauchamp, Kimberlee Chambers, Melissa Belfry, and Tamara Koltermann-Hernandez.

Suggestions from anonymous reviewers at the proposal stage have enhanced the quality of the project. Thank you for your insight and vision. Although I cannot name them, I also thank the feminists whom I asked to review material for this book. Their labor permitted me to undertake a formal review process for some of the authors' contributions. Blackwell has been wonderful throughout the whole process – from the proposal stage, through submission, to publication. Sarah Falkus, Michelle O'Connell, Joanna Pyke, Katherine Warren and Caroline Wilding have been particularly helpful! Thanks to Joan Sharpe who provided editorial assistance in preparation of the manuscript.

Numerous colleagues have supported this project in intangible, indirect, and unplanned ways: Margo Matwychuk, Margot Young, Martha McMahon, Radhika Desai, David Butz, Janice Monk, J. P. Jones, Lawrence Berg, Robin Roth, Shona Leybourne, Susan Ruddick, Sara McLafferty, Lynn Staeheli, Ellen Hansen, Ruth Liepins, Katherine Teghtsoonian, Marie Campbell, Michael Prince, Marge Reitsma-Street, and Anita Molzahn. I thank you all. I also acknowledge the Universities of Victoria and of Vienna from whom I received support during the course of the project.

Finally, I thank Karl, Ann, Kath, Clarice, and Kathy for creating a respite during undertaking this project. And I thank Karl for his enduring support of me and all my endeavors.

Figures and Tables

Figures

Table

Contributors

Elisabeth Bäschlin has taught human geography (urban geography, planning, and development) at the University of Berne, Switzerland, since 1978. As part of her research, she is interested in feminist geography, especially European farm women and women in liberation movements. She is a founder of the Working Circle for German-Speaking Feminists in Geography, was editor of the *Geo-Rundbrief*, and is also responsible for a small NGO working with people from Western Sahara.

Lawrence D. Berg teaches feminist social and cultural geography at Okanagan University College, Canada. He has published a number of articles on gendered geographies.

Liz Bondi is professor of social geography at the University of Edinburgh, UK, where she teaches on the geography program and on the gender studies program. Her research interests include counseling and psychotherapy, gender and urban change, and theorizations of identity, self, and subjectivity. Her work draws strongly on debates in feminist theory.

David Butz is associate professor of geography at Brock University, Canada. His interests in social and cultural geography, geographies of everyday resistance, community level social organization in northern Pakistan, and transport labor in the Karakoram/Himalaya coalesce in a research project which investigates how portering relations have significantly shaped transcultural interactions in the Karakoram region. He is also gathering courage to write about "exodic" (exodus-focused) spatial identities as expressed in reggae music and Rastafarianism.

Meghan Cope is an assistant professor of geography at the State University of New York at Buffalo, USA, where she is engaged in research and teaching on urban problems; spatial perspectives on the social constructions of gender, race, and class; geographies of social welfare, education, and housing; and methodological issues in doing qualitative research.

Isabel Dyck is a social geographer in the School of Rehabilitation Sciences and Faculty Associate, Women's Studies, at the University of British Columbia, Canada. She teaches qualitative methodology, geographies of disability, and feminist theories of the body. Her research interests include women with chronic illness, and resettlement experiences of family, school, and health care.

Kim V. L. England teaches feminist and urban geographies in the Department of Geography at the University of Washington, Seattle, USA. She is also Adjunct Professor of Geography at the University of Toronto, Canada. Her research interests revolve around feminist theories, women's paid work, service sector employment, child care, and families.

Karen Falconer Al-Hindi teaches introductory women's studies and human geography, as well as upper-level and graduate feminist geography, at the University of Nebraska at Omaha, USA. Her research includes the history and philosophy of geography, feminist research methods, urban planning, and feminism, ethics, and geography.

Feminist Pedagogy Working Group in Victoria, BC, Canada: Kathleen Gabelmann, Jenny Kerber, Andrea Lloyd, Pamela Moss, Denise Pritchard, Amy Zidulka, Joy Beauchamp, Melissa Belfry, Kimberlee Chambers, Tamara Koltermann-Hernandez, Erin Quigley, Cristal Scheer. We came together with a shared interest in feminist research in geography and a desire to explore the relationship between theory and practice. Over the course of nearly a year, we discussed our diverse experiences as feminists, as students, and as researchers. We agreed to produce material that supports feminist research pedagogy. Sharing an enthusiasm about working through the challenges and possibilities of feminist research, we drew upon both the contributors' and our own questions, dilemmas, and ideas about how to take on, think about, and do feminist research in geography. In our contribution, we provide a glimpse into our own process of working through the material in the book. We hope that our work, while not intended to be prescriptive, promotes discussion and enriches your learning.

Mary Gilmartin is a doctoral candidate in the Department of Geography at the University of Kentucky, USA. She is currently completing her dissertation, a comparative study of education during political transition in Northern Ireland and South Africa.

Louise C. Johnson is a feminist geographer currently Head of the art school at Deakin University, Melbourne, Australia. In this role she is juggling management with research but also actively engaging with the "cultural turn" in geography by studying urban festivals and cultural capital. In addition she is researching women workers in the information technology and service industries and on kitchens.

Hope Kawabata was a graduate student at the University of Nebraska at Omaha, United States between 1995 and 1997. Her thesis was on ethnic minority women in Omaha who worked from home by using telecommunication technologies. Her research focused on the relationship among gender, race, ethnicity, and space in telecommuting. She would like to convey to students reading this text that coming forth and writing about her experiences while doing an MA has been worthwhile!

Mei-Po Kwan is an associate professor of geography at the Ohio State University, USA. She is also a consulting editor of *Geographical Analysis*. Her research interests include women's spatial mobility, information technologies and women's everyday lives, and feminist methods.

Deirdre McKay is currently at the Australian National University as a postdoctoral fellow. With research collaborators from the Philippine Women Center, she is working on patterns of female migration and transnational feminine subjectivities.

Joan Marshall teaches environmental studies at McGill University, Canada. Her research interests focus on social and institutional change, especially in relation to youth and women. Her experience with a long history of community activism reflects a personal commitment that continues to pose dilemmas in carrying out her research.

Pamela Moss teaches feminist methodologies in the Studies in Policy and Practice Program at the University of Victoria, Canada. Her research interests include body politics, feminist theory, autobiography, and women's experiences of chronic illness. She is also a community activist involved with creating innovative housing programs for women in crisis.

Karen Nairn is a research fellow at the Children's Issues Centre, University of Otago, New Zealand. Drawing on her geography and high school teaching background, Karen is researching the spatial and social dimensions of young people's lives in schools and public spaces. Other research interests include feminist theory, embodiment, geographic knowledges and identities, and the politics of voice.

Geraldine Pratt is professor of geography at the University of British Columbia, Canada. She is editor of *Environment and Planning D: Society and Space*, co-editor of the *Dictionary of Human Geography*, 4th edition, and co-author of *Gender, Work and Space*. She works on gender, race and labor markets, and theater (and focus groups!) as methodology and politics.

Maureen G. Reed tries to understand policy processes associated with environmental management. She was a professor at the University of British Columbia for almost ten years before accepting a position at the University of Saskatchewan in July 2000. Her research interests have taken her across most of Canada, where she has explored the social character and political responses of rural communities that are facing changes in the non-human environment and in government policy agendas.

Gill Valentine is professor of geography at the University of Sheffield, UK, where she teaches social geography, approaches to human geography and qualitative methods. She is author of *Social Geographies*, *Consuming Geographies*, and co-editor of *Children's Geographies*, *Cool Places* and *Mapping Desire*.

1

Taking on, Thinking about, and Doing Feminist Research in Geography

Pamela Moss

What makes research in geography *feminist*? If you're a feminist, do you *have to* do feminist research? And, if you're *not* a feminist, can you do *feminist* research? What sorts of things do you *need* to know about in order to *do* feminist research? How do you go about *making* a conventional method *feminist*? Can the *practice* of geography research actually *ever* be feminist?

I began thinking about this book with this seemingly endless list of questions in mind, questions with definitive answers nowhere in sight. Then, I wondered about wanting definitive answers. I thought what a treat it would be to know when I had one and how suitably impressed I would be when I saw one. Yet I'm content not to know. In fact, I revel in not knowing – not knowing for sure. I'm comfortable asking questions about research that have no "right" answers, to talk endlessly about how feminism influences research in geography with whomever has similar inclinations. I've been interested in feminist geography research for what seems like ages now, as an undergraduate stealing glances of *Antipode* for special projects, research papers, and for any chance I could get. I eventually figured out that the path to being an academic – studying, obtaining degrees, and landing a tenure-track position at a university – seemed to be a worthwhile path to follow so that I could continue being a feminist while being employed. All these years later, after having undertaken various types of feminist research projects in geography and teaching feminist methodologies in a number of contexts, I decided that I wanted to pull together a collection of works that was organized around issues that I found useful in undertaking feminist research in geography. For me, and I would anticipate that for others this might also be the case, it makes sense to sort feminist research into processes that we engage in when putting feminist geography into practice: taking on, thinking about, and doing feminist research. And, to be sure, these processes only make sense in the context of the history of methodological work within feminist geography.

Even though developing a feminist analysis was an issue early on in the radical movement in geography, methodological concerns began appearing in print only in the 1990s (see for example McDowell, 1992a, 1993a, 1993b; *Canadian Geographer*, 1993). It wasn't that feminists in geography weren't interested in doing feminist research; rather, feminists weren't publishing their thoughts on feminist methodologies. It soon became important however to refine feminist concepts in geography, including those concepts associated with doing feminist research – method, methodology, and epistemology (Moss, 1993, pp. 48–9). These early methodological works were heavily influenced by feminist work done in the early and mid-1980s (see for example Moraga and Anzaldúa, 1981; Roberts, 1981; Bowles and Klein, 1983; Harding, 1986, 1987a; Hartsock, 1984). In fact, Sandra Harding's (1987b, pp. 2–3) definitions of *method* as techniques used in gathering evidence, *methodology* as a theory and analysis of how research should proceed, and *epistemology* as a theory of knowledge, are still powerful beginning points in understanding processes involved in undertaking feminist research. As debates unfolded within and outside geography throughout the 1980s and 1990s, feminists worked out more sophisticated definitions, especially as they related to racialized and sexualized relations within feminist scholarship (see for example Sedgwick, 1990; Mohanty, 1991; Collins, 1998). The crux of these concepts remained the same – method has to do with *doing* research, methodology had to do with *approaching* research, and epistemology had to do with *knowledge associated with doing and approaching research.*

Attention to methodological issues in feminist geography coincided with the increased publication of debates in collections of works focusing on a specific aspect of feminist methodology in women's studies, sociology, and anthropology (see for example Personal Narratives Group, 1989; Nielson, 1990; Fonow and Cook, 1991; Gluck and Patai, 1991) and of more generalized handbooks or "how-to" books (see for example Eichler, 1988; Kirby and McKenna, 1989; Smith, 1990a, 1990b; Reinharz, 1992). In geography, these feminist debates manifested in collections of journal articles (see *Canadian Geographer*, 1993; *Professional Geographer*, 1994, 1995; *Antipode*, 1995), sections of books on feminist geography (see Jones, Nast and Roberts, 1997a; WGSG, 1997; McDowell, 1999), and single articles appearing in wide variety of feminist and non-feminist geography journals (see for example Pratt, 1993, 2000; Katz, 1996; Moss and Matwychuk, 1996, 2000; Domosh, 1997; Rose, 1997; Nairn, 1999).

This interest in methodology among feminist geographers was not only a part of how feminism shapes feminist research in geography but also, as Susan Hanson (1997, p. 122) points out, part of how geography shapes feminist approaches to research. Feminist geographers took up topics that were specific to the discipline: spatializing the constitution of identities, contextualizing meanings of places in relation to gender, and demonstrating how gender as a social construction intersects with other socially constructed categories within particular spatialities, among many other topics. Being able to work through these types of interests has had an impact on the way feminists approach research within geography ranging from approaching research as a feminist, through integrating spatial conceptualizations into a feminist research framework, to choosing feminist methods for collecting and analyzing information. The maturity of the methodological arguments developed by so many feminists within the past several years makes feminist geography a rich field from which to draw out specific research practices. Being a feminist matters when taking on research in geography in that a feminist politics – whether it be based on pro-woman, anti-oppression, or based on social justice – influences all aspects of the research process. Thinking about feminist research tends to sharpen an approach to a project in that understanding power and knowledge brings into focus the varied contexts within which research takes place. Doing feminist research means actually undertaking the task of collecting and analyzing information while engaging a feminist politics. By including pieces written by different feminists with different perspectives on research and methods, I am able to offer a collage of ideas, thoughts, and arguments about the *practice* of feminist geography. Instead of reiterating the arguments about method, methodology, and epistemology by way of introducing these works, I turn the kaleidoscope just a bit and focus on sets of issues that have arisen out of those discussions. As a way to make my way through these issues, I first discuss taking on, thinking about, and doing feminist research in turn and through the discussion offer a possible framework for understanding specific practices in feminist research in geography.

Taking on Feminist Research

Taking on feminist research entails close scrutiny and (re)politicization of all aspects of the research process – from choosing a research topic to selecting data collection methods, from setting a research question to conceptualizing theoretical constructs, and from designing a research project to presenting and circulating analyses. Working with the variegated contours of the infusions, interfaces, and articulations of feminism and research is a first step in taking on feminist research in geography. Placing feminist work as well as placing yourself as a feminist researcher in the context of research in geography and in feminism – contextualizing your work – makes it easier to see where you are coming from and where you see your work going.

Though perhaps tiresome to both ask and answer, being able to figure out why a piece of research is feminist continues to be important. Feminism has often been differentiated by distinguishing waves of political approaches to explaining and understanding women's lives. "First wave" feminism is associated with social reform, suffrage, and temperance movements; "second wave" with equitable pay, sexual liberation, and consciousness-raising; and "third wave" with difference, speaking from the margins, and positioning self and other within multiple oppressions. And, now, as we are moving through the new decade of the twenty-first century, feminism is being reconstituted into *feminisms*, ones that go beyond gender as the central construct in defining any feminism (see for example Hekman, 1999; Oakley, 2000), beyond power conceived dichotomously as either something to hold or something to be used (see for example Collins, 1998; Sandoval, 2000), and beyond body as the home and/or conduit of being and experience (see for example Kruks, 2000). With the increase in various influences affecting the constitution of feminisms, it becomes more and more difficult to differentiate pieces of work that use feminist frameworks, feminist theories, or feminist constructs to provide critical or radical readings, research, and analyses and those that are indeed feminist. At the risk of being essentialist, that is promoting the idea that there is a feminist *essence* that exists in *all* feminist research, I think that it is useful to unravel explicitly the ties that bind a piece of research in geography to a particular feminist politics, a particular feminism. Refusing to accept that there is one singular feminist politics does not preclude identifying straightforwardly how an author of a research text is engaging feminism in the sense of not only abstract concepts, but also concrete actions.

Being able to scrutinize more closely the ways in which we take on feminisms in research may be a way to open up debate with non-feminists as well as among feminists themselves. With non-feminists, debates could take up the issue of what advantages do feminisms offer researchers that non-feminist research can't and, perhaps, vice versa. Unfortunately, what happens in this type of debate is that the potential overlap of views that is the basis for exchanging ideas is quite limited and therefore falls flat as many feminist geographers no doubt have experienced in classrooms, conferences, and colloquia. "Opening up debate" among feminists has its own set of problems. In an academic milieu that is masculinist in its practices, how can feminists wholly resist reproducing these practices and remain *feminists* and *academic researchers*? Much feminist research in geography is masculinist in its practice, not out of intention, but moreso out of training for being an academic and for survival in the field. Throughout the research project, feminists are continually holding in tension the immediacy of constructing authority through doing research, writing about it, and teaching it and the notion that what they are doing initially emerged as a contestation of an existing orthodoxy. Paradoxically, even while negotiating this tension, within writings about feminist research in geography there has arisen seemingly inevitably a feminist orthodoxy in the English-speaking academy, one that tends to value qualitative research and reflexivity as cornerstones of feminist research in geography. Paying attention to the wide-ranging and perhaps roaming definitions of feminism that infuse feminist research in geography and then engaging with those ideas with feminists in discussion and in print could possibly release some of the tension and facilitate a way through this dominance toward opening up what it might mean to take on feminist research.

How this debate takes place and the form it inhabits is open. I think that this book is one attempt at trying to rupture the closely knitted visions of feminist methodologies in geography and to rumple the smooth progress of developing decidedly feminist approaches to research in geography and accentuate the highly contingent performance of feminisms in feminist geographers' works. If taking on feminism in doing research in geography makes a difference, then learning about how feminists have come to a feminism in their work is useful. The content of each piece has a particular relationship with methodology, epistemology, and method. Elisabeth Bäschlin provides a brief history of the forging of feminist geography in German-speaking countries. Her tale plays out in four scenes with pioneering feminists weaving networks and eventually entering institutions so as to shape more fully the future of taking on feminism in geographic research. Mary Gilmartin reflects on her personal journey toward geography through her readings of Toni Morrison. She comes to understand that she can access experiences in ways that will assist in negotiating the tension between knowing, learning, and doing. Meghan Cope lays out what types of feminist claims about knowledge affect the undertaking of research. She unravels just one type of bind that marks a piece of research as feminist. Louise Johnson draws on her own experience with a research project with women looking for employment. She recounts how feminism makes a difference in research activities including securing funding, hiring research assistants, and analyzing data. Bearing in mind the contingent ways feminisms articulate in specific research projects, trying to identify connections among feminisms, geographies, and research while reading these contributions is but one entry point into understanding what taking on feminist research involves.

Thinking about Feminist Research

Issues arising when thinking about feminist research, though similar, manifest differently than when taking on feminist research. The extent to which feminisms influence research processes as well as the translation of feminist politics into research are just as significant. Yet thinking about feminist research also includes the articulation of specific theories *with* a feminist methodological approach. For example, being able to interweave thoughts about identity, subjectivity, and self requires thinking through how to access salient information as a feminist as well as how to create a feminist framework for understanding identity, subjectivity, and self. Through this process, ambiguities, contradictions, and paradoxes emerge alongside relative certainties, congruencies, and consistencies about both the content and the process (methodology). These seemingly opposing empirical findings permit researchers to continue pursuing ideas, thoughts, and notions about the topic and how to do research. Not all geographers undertaking feminist research focus on identity, subjectivity, and self as research topics; however, these topics have been important in understanding the relationships researchers have with themselves, research partici- pants, research topics and thus have come to play a large part in understanding feminist methodology (see for example McDowell, 1993b, 1997b; Nast, 1994). The particularities of methodological discussion regarding topics, themes, and the manner of engagement are specific to the feminists thinking about research. Within feminist geography, researchers have tended to think about power, knowledge, and contexts together with sorting out the practicalities of doing research.

Power is a central construct in discussions of how to approach feminist research and differences in conceptualizations of power produce different types of feminist approaches to research. Feminist methodological discus- sions rarely revolve around competing conceptualizations of power that would be useful for feminist research; rather, discussions of power usually promote one particular conceptualization. For example, Gillian Rose (1997) argues that feminists who discuss "distributions of power" invoke a structuralist account of power that is not particularly useful for feminists. She claims that using the notion of uneven landscapes of power refuses to acknowledge that people variously located in complex webs of power participate in their own constitution. As a result of invoking such a transparent notion of power, feminist geographers only end up creating a transparent reflexivity – something that is impossible to achieve because no one can know themselves thoroughly and exhaustively. In contrast, Linda McDowell (1997b), dealing with the same sets of issues – destabilization of the category woman, what makes research feminist, and gender – comes to a different conclusion. Even with the shift in focus on *how* gender is constituted through power, she remains focused on transformative understandings of women's conditions of everyday life. She makes the point that holding onto notions of power that conceive social relations as flexible but not too flexible permits complex abstractions to explain more adequately complexities in everyday life. What is interesting about these two methodological discussions is that they both focus on identity and difference – of the researcher and of the research participant. These same interests reproduce feminist research in geography differently in different contexts (for example see Peace and Longhurst, 1997, for the Aotearoa/New Zealand case; Bäschlin and Meier, 1995, for feminist geography in the German-speaking academy). In disciplines other than geography, feminist researchers discuss approaches to research and power outside identity and difference. For example, in North American sociology, struggling to justify qualitative methods in light of quantitative dominance shapes feminist discussions of methodology (for example see Devault, 1999; *Resources for Feminist Research/Documentation sur la recherche féministe* 2000) whereas in North American psychology, unsettling links among masculine power and subjectivity fashions discussion (see for example Swann, 1997; Ussher, 1997).

Feminists undertaking research in geography think about multiple aspects of knowledge through an array of relevant concerns. Geography as a discipline has privileged a masculine subject position and reproduced binaries such as male/female, culture/nature, and object/subject where more value has been placed on the first part of the dichotomy. Feminists in geography have followed the lead of several feminist philosophers in examining the underlying assumptions of who are knowers, what can be known, and what is valued as knowable (see for example Hawkesworth, 1990; Harding, 1991; Haraway, 1991; Tuana, 1993). Understanding how masculinity permeates the discipline has opened up ways of thinking about knowledge such that a feminist subject positioning can develop within geography as well as that the same, masculine-weighted binaries are not continually being reproduced. In coming to terms with these sorts of assumptions, feminists also distinguish processes that construct or constitute knowledge – processes that are primarily discursive such as reiterating masculine words (for example, mankind), concepts (for example, objectivity), and notions (for example, exploring, conquering, and subduing the exotic as integral to the practice of geography) and processes that are primarily material such as mentoring students, training researchers, and teaching students (for examples of these types of arguments see Berg and Kearns, 1996; Blunt and Rose, 1994; Desbiens, 1999; Moss et al., 1999; Hanson, 2000). Because these processes are saturated with power even within feminism, a politics surrounding the construction of knowledge endures beyond the overt actions involving written or spoken words. More subtle activities, such as the choice of authors in a reading list for a senior undergraduate class, of a book to be reviewed in a journal, or of a cited work as an exemplar of a point a scholar wants to make continue (re)creating an authoritative knowledge that may or may not challenge the dominant orthodoxy in feminism. Being able to read works critically under the conditions within which one learns implies untangling the processes constructing that specific contribution as a piece of knowledge as well as part the process of creating knowledge that would include that specific contribution.

In addition to issues involving power and knowledge, thinking about feminist research entails thinking about the context within which research takes place. Because power is intimately tied up with the construction, constitution, and production of knowledge through research, the context within which research can take place also needs close inspection. Take, for instance, funding and time, two of the most limiting and enabling aspects of research. Ample time and money creates an environment where research can actually take place. Yet having both does not necessarily entail an unproblematic research process. Questions immediately arise as to whether to accept money from, for example, a corporate entity, a philanthropic foundation, or the state or to hire research assistants to increase research time for the project. The latter of course further begs the question, what sorts of employment relationships are part of feminist research? Designing research projects sensitive to notions of power and knowledge takes a considerable amount of planning. Issues for thought range from, for example, "appropriate" attire to seat location while conducting interviews; from etiquette for contacting potential research participants to remuneration of actual participants; and from facilitating relationships among research team members and participants to enabling a supportive environment for training research associates in the field. Dilemmas inevitably emerge even with careful, thoughtful, and thorough planning and not all quandaries can be resolved – immediately or in the long term. The context of research also includes understanding issues beyond the immediacy of undertaking a feminist research project. For example, in order to secure funding, researchers need to figure out what types of research agendas are being advanced by particular funding agencies so that applications for funding are directed to appropriate institutions. Also, recognizing conventional practices of the academy *in specific places* is important so that an aspiring feminist researcher knows whether to engage in local struggles over justifying feminist research as "legitimate." Thinking about research in the context of feminist research then includes understanding the specificity of the spatialities of both the research process and the milieu of feminist research.

The authors in this book have given thought to and written about specific aspects of power, knowledge, and context, either explicitly or implicitly. Liz Bondi explores a paradoxical space within feminism. She relates her experiences with the journal *Gender, Place and Culture* as an example of a feminist politics in an uncertain space as part of the context within which feminist geography contributes to creating knowledge. David Butz and Lawrence Berg present some of their thoughts on being male and trying to do feminist research while working through notions of masculine dominance in the politics of knowledge production. They offer an innovative conceptualization, a duppy (which refers to a variety of sly and malevolent ghosts) feminist, that may describe more sincerely the power dimensions among men engaging feminism. Karen Falconer Al-Hindi and Hope Kawabata use Hope's research project as a backdrop against which Karen argues that feminist researchers do have the potential to be more fully self-reflexive in the pursuit of understanding power relations in the context of interviews. Unlike some of the prevailing understandings of reflexivity in feminist geography, Karen claims that an equitable power relationship between researchers and research participants is possible. Gill Valentine recalls some of the situations in her research projects that call into question assumptions about sameness and difference. She argues that her performance of her gender and sexuality is context-specific – varying from project to project, interview to interview precisely because negotiations and readings of both are momentary and specific. From these contributions, it becomes more feasible to think that sorting through issues of power, knowledge, and context may pave the way for actually doing feminist research.

Doing Feminist Research

Issues surrounding doing feminist research in geography are in a sense an amalgamation of matters arising when taking on and thinking about feminist research. Paying close attention to how ideas about feminism, power, knowledge, and context play out when undertaking the research itself and engaging particular research methods are part and parcel to doing feminist research. Without a continuation of thinking through these issues, the work going into designing a *feminist* research project might be lost.

Three key concerns shaping the doing of feminist research in geography are the scales of analysis and project, analytical issues emerging from engaging in the research process, and the choice of data collection method. The scale of analysis – the spatial focus of the inquiry – differs from the scale of the project – the spatial extent of the research. Feminist research can be undertaken on a variety of scales – for example, local, regional, national, and international – with a variety of scales of analysis – for example, body, people, home, institution, city, or region. Although feminist research often focuses on local, micro-scale studies, there is no intrinsic connection between feminist research and scale. Drawing out the implications of a specific scale of analysis as well as the scale of the project for the topic at hand is part of what research in geography is about (see the organization of McDowell's 1999 book).

What appear to be also significant in doing feminist research are the analytical issues emerging from engaging in the research process itself. Questions concerning the implementation of a feminist epistemology and feminist methodology manifest during the undertaking of research and emerge as problems or dilemmas. Sometimes analytical problems can involve incongruent knowledge claims, as for example, arguing that the experience of marginalized women is the (only) basis for political action while using a structuralist framework that situates experience outside accessible knowledge. Or, perhaps problems have to do with the incompatibility of topic and theory, as for example, focusing on detailed social practices of an institution and empowerment without accounting for the notion of how power is deployed through social relations of power in institutional settings. Both these types of problems cause difficulties in providing an analysis that makes sense. Problems may also arise when the methodological approach of a research project is at odds with the chosen theoretical framework. For example, maintaining a complex conceptualization of power methodologically throughout data collection and analysis (through a specific understanding of identity as fixed) while theoretically challenging the same conceptualization of power in the explanation of the phenomenon (identity as changing) can be difficult. Undesired slippage between concepts is common and can usually be identified and dealt with through discussions with colleagues, exchanges of works in progress, and write ups of the research. Addressing these types of problems as they emerge can strengthen the analysis of the topic of a research project as well as refine the methodological approach.

With regard to the choice of method in feminist geography there doesn't seem to be a question as to whether feminists "should" be using qualitative or quantitative approaches either for data collection or analysis (for an extensive exchange of ideas see *Professional Geographer*, 1995). Rather, the predominant view seems to be choosing a method *appropriate* to the research question. Feminists have argued that the issues brought to the fore during the challenge to quantitative methods, especially the exploitative nature of the relationship between numbers and people, are moot in the sense that qualitative methods can be just as exploitative (see for example the argument in Stacey 1988). Calling for more sensitivity to the relationship between the researcher and the research participant (often referred to as "the researched") definitely heightened awareness about the actual choice of method for data collection and analysis. Interviewing women, for example, was not just about interviewing women (in contrast to Oakley's 1980 argument), and the types of conclusions that one could draw when basing the entire collection of information on the notion that women had some connection to each other because they were women had to be questioned. Where that woman was located *vis-à-vis* the multiplicity of power relations mattered when it came to interacting and deploying power within the research process. Interviewing elite women, for example, was different, and perhaps could and should be approached differently than interviewing women marginalized by the same economic processes that made the first woman a member of an elite. These choices of method, too, are inexorably shaped by the types of questions feminist researchers in geography are interested in asking. Certain methods seem to be associated with certain kinds of research – ethnography with cross-cultural research, focus groups with minority groups being studied by majority groups, and autobiography with marginalized women. But this is not always the case. Ethnography can be used within one's own culture; focus groups can be used as a way of studying "us" instead of "them"; and autobiography might be useful in addressing privilege.

The authors writing about doing feminist research discuss specific research projects in terms of project design, choice of method, and dilemmas surfacing once the research has begun. They also address issues concerning scales of analysis and projects, analytical issues emerging while engaging in the research process, and choice of methods in their chapters. Maureen Reed teases out fibers that hold in balance the "needs" of the researcher and the "demands" of funding agencies. She uses examples from several forestry research projects with different scales of analysis and shows how this tension can be balanced. Karen Nairn pulls together some of her thoughts about conducting multi-method research. She politicizes the notion of "field trip" in the vein of politicizing "fieldwork" and ends up exposing processes that construct geographic knowledge. Mei-Po Kwan argues that understanding quantitative analysis is important for feminists so that they can root out masculine bias. She works through specific examples and pulls out the epistemological claims that make the research either feminist or not. Joan Marshall presents some of her deliberations over choices she has had to make while negotiating personal and professional relationships with the people in the community where she is undertaking her research. Because the community she is studying is so small and replete with complex social relations, she must continually scrutinize and assess her interpretations and presentations of information from her research. Deirdre McKay seeks to problematize the personal interaction between researchers and research participants: she is both enthusiastic and reluctant to disclose information about both herself and the women she talked with. She suggests that resolution may come through critical analysis of not just the topic, but the process of engagement. Kim England presents detailed examples of her experiences of interviewing elite women. She draws out the dissension between her expectations and the actual happenings of the interview setting and provides useful tips on how to adjust. Geraldine Pratt raises questions about the process of interpreting material collected through focus groups. Through her critical reading of printed transcripts, she encourages feminists to access the potential focus groups have to offer in providing insight into topics such as power and identity. Although these contributions are not exhaustive of the types of feminist research geographers undertake – some obvious omissions are institutional ethnography, survey research, and participatory action research – they do represent the range of methods taken up by geographers doing feminist research. These contributors demonstrate *en masse* that in doing feminist research, taking on feminism as a methodological approach to research matters and that the thinking about feminist research doesn't stop.

Feminist Methodologies in Geography

To reiterate, feminist methodology is about the approach to research, including conventional aspects of research – the design, the data collection, the analysis, and the circulation of information – and the lesser acknowledged aspects of conventional research – relationships among people involved in the research process, the actual conduct of the research, and process through which the research comes to be undertaken and completed. But it is not just adjustments in the definition of methodology that make a methodology "feminist." Making a methodology "feminist" implies politicizing a methodology *through* feminism. As we already know, a feminist politics has a wide range of possibilities. These possibilities are realized through our own translations of a politics into practice (read praxis). Our exchanges of information, experience, and knowledge through various types of interactions – taking and teaching courses, attending and giving workshops, giving and receiving advice, writing and reading papers – seek to further refine existing arguments as well as open up new spaces for new, innovative directions for future feminist research in geography. Whether it be in formal lectures or informal conversations, critiques are ongoing and are necessarily part of learning, understanding, and engaging feminist research.

What may be helpful in figuring out how to approach, assess, or affirm feminist approaches to research in geography is to contribute to discussions of how feminist methodologies play out in the politics of doing research in and out of the classroom. I have found three discussion points, conceived as tensions rather than polarities, useful in igniting dialogue.

Feminism as a politics is sometimes difficult to grasp for those not already committed. The increased incidence of younger women refusing to claim to be feminists might be directly related to the backlash against the public gains feminists have made in the last quarter of the twentieth century and the negative media images of individual feminists (see Faludi 1992 for a popular take on this issue). Is the question really about whether or not you call yourself feminist, or is it about taking up the politics associated with a particular feminism? Or, can you call yourself a feminist without being politically committed in your research or in your daily life? And, if you're not committed, are you being a dilettante, perhaps shopping for a politically correct stance for future job opportunities, and is that a "bad" thing? Coming to terms with this tension between political commitment and dilettantism is embedded in our everyday existence and can be deeply troubling.

As can the tension between theory and praxis. Theory, as a combination of both conceptualizations of phenomena and an explanation of how phenomena work, exist, or articulate, and praxis, as a politically active way to live in the world, are undeniably linked. Understanding one as integrally wrapped up within the other creates an environment where there cannot be any act that is not political. Understanding them as separate entities permits neatly carved out niches among those interested in theorizing life (for example, academics) and dealing with social injustices (for example, activists). Is either a solution? Many feminists hearken back to Karl Marx's words, and point out that the contribution of feminism is not (only) to provide an understanding of the world, the point is (also) to change it. Yet living daily life always being politically engaged is emotionally painful and ethically debilitating. Are there guidelines to resolving such a tension for feminists? For feminist researchers?

When disagreements emerge over things like the extent or intensity of political commitment and the practice of theory, how, when, and in what context do feminists express criticism? Being aware of the tension between maintaining solidarity and engaging in critique is crucial in practicing a feminist politics. What is difficult in negotiating this tension is to "know" when to support other feminists, even though you disagree with them, and when to speak out against feminist actions, even though you might support the end result. In the context of feminist research in geography, this may play out in a variety of ways. You might give support to feminists presenting their work in departmental colloquiums even though you adhere to an opposing theoretical framework whereas you might choose to publish a critique of the same feminist's work in a feminist journal, perhaps outside the reach of the department's immediate attention.

The outcome of discussing these tensions in the context of taking on, thinking about, and doing feminist research is uncertain. It may polarize discussion between feminists and non-feminists. It may pull feminists apart and set up a continuum of "pure" and "tainted" feminist politics. Or, it may, as it has in my experience, evoke a set of ideas that establish a fresh, context-specific framework for engaging methodological debates in feminist geography. The process of setting up the framework through discussion might set a collective, engaging tone for reading, discussing, and critiquing the contributions in this textbook.

About the Book

I set the purpose of this book as threefold. First, I wanted to put together a textbook with a wide variety of feminist perspectives on putting feminist geography into practice, or how to approach research in geography as a feminist and how to undertake feminist research in geography. Several influences within feminism are represented in these chapters: environmentalism, Marxism, poststructuralism, postmodernism, and socialism. There is also a noticeable difference between second and third wave feminisms as well as between feminisms that deal primarily with discourses and those that deal with materialities. What is profoundly attractive about all these contributions is that they are all "feminist." Although all may not echo your particular feminism or feminist stance, my hope is that some of the work at least resonates with your experience or piques your interest. Second, I intended to capture a *sample* of leading feminist research from a variety of feminist geographers. The contributors have various relationships with the English-speaking academy with between three and over twenty years of experience as feminist geographers – ranging from a complete "outsider" to a well-ensconced "insider," from undergraduate students to full tenured professors, from the "margins" of Australia and New Zealand in the English-speaking academy to the bi-nodal "center" of North America and Britain. Locations of these contributors within the complexly spun web of power relations and social divisions vary according to sex, age, gender, class background, expressions of sexuality, race, ability, and ethnicity. Access to such difference may or may not be easy for the reader unless that difference is disclosed as part of the discussion about methodology. Third, I aimed to create a collection that participated more fully in demystifying the research process and making research accessible in various ways. Rather than portraying research as something too important, too complex, or too difficult for women and feminists to undertake (unfortunately, a still too common belief!), I sought to unravel, in bits and pieces, the research process by inviting other feminists to write specifically about one particular aspect that I thought might interest them. Breaking down research into parts makes the tasks not only "do-able," but also "identifiable" – not in a search to simplify research, but rather to make each aspect more comprehensible. By piecing together the text, I was able to highlight what I thought important for discussion: rather than focusing on rigor, validity, reliability, and bias (points upon which non-feminists attempt to debunk feminist research), the book tends to focus on personal and political struggles, rethinking research strategies, and embracing contradictions.

With such a tightly woven purpose, one could walk away with the notion that this was the intent from the beginning. I'd rather not have that happen. This book has its own history, one that is only partly located in feminist methodologies in geography. Originally, I had planned a feminist book on autobiography and many of the contributors had agreed to write about their experiences of being a feminist geographer. At the book proposal review process, however, the book transformed into a book on feminist methodologies in geography. Yet even this process was subject to revisions – initially the book was conceived as a text on approaching gender and geography, but as the contributors created their pieces, it became clear that this was a book on feminist geography and not gender and geography. All the contributors remained committed to the project, however, and the change in the final list of contributors reflects competing commitments more than intellectual differences. Some contributions went through a formal peer-review process at the submission stage, whereby I was primarily the editor. I acted as both an editor and a reviewer for almost all the contributions. Through the editing process, I strove to create a tone that was personal, informal, erudite, and critical in hopes of producing an accessible text for undergraduates and new graduate students. Even though I saw myself as being picky beyond what I usually can muster enough in courage to display, all authors eagerly took on this challenge, and succeeded.

In organizing these contributions, it seemed to me that the themes of "taking on," "thinking about," and "doing" feminist research in geography encompassed what it was that I thought was needed in a textbook on feminist methodologies in geography. Taking on feminism in research is a political act. Yet being a feminist in geography is not necessarily difficult, nor is it necessarily simple. Depending on the way feminist research in geography is situated within the most immediate institutional environment for geography (department, university, for example) and the way the discipline is conceived, presented, or "taught," the path for developing as a feminist in geography or contributing to feminist geography will vary. Although uneven in constancy, feminism still struggles for legitimacy within geography and the academy. Thinking about feminist research is crucial to being both a feminist and a researcher. What feminists decide to discuss about research stems directly from the meeting point of intellectual moments, such as the focus on self, subjectivity, identity, and difference as well as power and knowledge, and their values, ethics, and politics, such as social justice, equity, anti-oppression, and experience. How to go about achieving goals in feminist research is a tangible purpose for being a feminist geographer. The goals, of course, are set through the arduous process of self-reflection with collective voice and of finding a way through theory and praxis. Doing feminist research covers the nitty-gritty of the actual data collection and analysis. But even while doing the most mundane task, one needs to think about the method, the analysis, and the use of information or data; the audience, the participants, and members of the research team; and about how feminism articulates with the chosen methods, form of presentation, and circulation of information or findings.

Another large part of the preparation of the book was the development of the pedagogical material, located at the beginning and end of each section. The Feminist Pedagogy Working Group, of which I was a member, consisted primarily of women undergraduate and graduate students in Victoria, British Columbia, who had some interest in feminism and geography. At one point, the group tried to include undergraduates at different universities through email connections, but this proved to be impractical. Each working member had answered a public invitation to attend a meeting about "putting together a textbook on feminist methodologies in geography." Not all members were feminist geographers, two were in English; not all were students, two had already graduated and three finished their studies during the project; and not all were the same age, the age span was nearly 30 years. Each woman was interested in drawing on their experiences, especially in the classroom, to develop material that would assist in making research more accessible to students like themselves. We held meetings where we discussed the content of each chapter in detail, possible questions that would provoke engaged discussion, and potential exercises that might enhance or challenge the point being made by the author(s). Three matters shaping the interpretation of the material in the chapters continually arose in the discussions – the definition of feminism, the use of language, and the creation of authority. The Group decided to convey the substance of these discussions as short essays that introduce each of the three sections of the book. In "Defining Feminism?" group members point to the variety of intellectual and experiential elements that have left impressions on their and other people's notions of what feminism is and can be. In "Delimiting Language?" group members question to what extent the use of jargon or precise language can be useful in the practice of feminist research. In "Decentering Authority!" group members draw attention to underlying thoughts about the processes through which authors forge, reproduce, sustain, and decenter authority. Communicating the nuanced meanings and the extensive array of discussion in written form has been taxing, for not all points can be represented adequately, and burdensome, in that choices have had to be made. The final form the pedagogical material takes was the most effective way the Group could express their engagement with the material. Members of the Group pored over each word, each question, and each exercise as a group, individually, and then as a group again. Group members offer each word, each question, and each exercise as *only one way* to engage the material presented in the chapters.

As with any writing project, especially textbooks, as the book took form, lacunae began to appear – some foreseen, others unanticipated. I knew that the contributors were all located in privileged and hegemonically powerful positions as members associated (at least at some time) with a university steeped in Western thought. I also knew that the topics addressed by the authors did not explicitly address racialization processes within feminist research, the problematization of the construction of 'race' and 'race' relations with feminist geography, or anti-racist strategies for effecting social and political change. More diversity along the lines of including feminist geographers from the South as well as topically would have only strengthened the collection. What I had not anticipated was the lack of variety in data collection methods and in analytical methods. Qualitative data collection methods dominate these pages and qualitative and textual analytical methods are by far the most popular types of analytical methods included in the book. But this lack of variety should not be too surprising given the propensity of feminist geographers in the English-speaking academy to reject quantitative methods as part of introducing feminism into geography. These elisions and omissions notwithstanding, I think that this book as a text will be an outstanding contribution to the practice of feminist geography.

Read, engage, learn, enjoy.

Part I

Taking on Feminist Research

Defining Feminism?

Feminist Pedagogy Working Group

Once "defined," feminism immediately becomes like an "it," a "thing," something graspable, something tangible. Not a process, a manner of asking questions, a way of looking at the world. Defining feminism is difficult because it's not something I have to articulate day to day. Of course, the question "*How do you define feminism?*" has different meanings for different women, different people. (A member of the Feminist Pedagogy Working Group)

Historically, both practically and academically, feminism has been part of a politics *for* women. Whether it be about control over reproductive rights, the construction of knowledge, or the concrete manifestations of patriarchal social relations, gender, defined as the social differences between males and females, was the central construct around which feminists developed theory and acted politically. During the last two decades of the twentieth century, feminism underwent two dramatic shifts. First, feminist critiques by women marginalized by feminist theory and within the women's movement shattered the category "woman," revealing it to be monolithic, one that was built on white, middle-class, Western women. Black women, lesbian women, disabled women, women from the South, and working-class women in their writings and in their actions showed that feminism was not for *all* women, a premise upon which second wave feminism was based; that type of feminism was only for an elite woman. Second, at the same time, this emphasis on difference among women coincided with the increased popularity of engaging with poststructural thought among feminist academics. An appealing aspect of poststructural thinking for feminists was the combination of destabilizing notions in the realms of knowledge and truth claims: rather than being singular, fixed, exhaustible, and universal, knowledge and truth claims were conceptualized as being multiplie, fluid, incomplete, and contingent.

> The resistance of many feminists to the "singular" – of woman, of politics, of gender – is what has attracted me to feminism and feminist writings. I understand feminism in its broadest sense as *praxis*, a form of political engagement aiming to analyse and to change inequitable and unjust social relations. It also is about exploring how to integrate personal feminisms in wider contexts, so that feminism's questions and goals do not just reflect the dominant interests of one 'kind' of woman. (A second member of the Feminist Pedagogy Working Group)

The *willingness* of feminisms to respond to critique and for feminists themselves to be self-critical means that feminism, or more accurately *feminisms*, has the potential to be an open and dynamic knowledge community. Yet there still exists a tension between feminists claiming a particular epistemological standpoint and claiming a legitimacy for multiple truth claims. While both positions advocate continual struggle against the elision and erasure of differences among women, among people, each does so in diverse ways.

> To me, feminism means accepting the notion that feminism may mean different things in different contexts. With my union, I'm always pointing out that there are issues that need to be addressed *for* women, *by* women. I'm the only one that knows anything about feminism! But when I'm with some of my feminist co-activists, I'm able to say something about marginalized groups within the city without having to be defensive or explain what I mean. And just because I say something about women in the street community, there is not the assumption that I think that they're the only marginalized group. Feminism plays out differently in each place, and I'm the one trying to figure out what feminism means and then trying to act on what I think is the best course of action. Still, feminism to me is about women somehow, while at the same time being about dismantling the injustices that systemic and personal abuse of power creates. (A third member of the Feminist Pedagogy Working Group)

This tension between a feminism that has as its primary starting point being a woman and one that relies on the destabilization of the category "woman" has been a catalyst for thinking through feminist research. Experiential knowledge is a cornerstone for second wave feminists, especially in a collective voice, just as understanding self, identity, and subjectivity is a cornerstone for third wave feminism. Drawing on women's experiences as a particular standpoint is a powerful way to demonstrate what it is about being a woman that is different than being a man: in childbirth, childrearing, and reproductive capacity, labor in the workplace and the home, and being in public and private spaces. Experience can also provide insight into new spaces, through which other truth claims can be made.

> Much of my motivation for going back to grad school was to figure out where I stood as a feminist. I had first come to feminism when, in my teens, it had helped me name and think through some of my own personal experiences. I figured, as many others did, that feminism was based on the premise that *all* women are linked through a common experience of oppression. By the time I was in my early twenties, my interest in feminism had waned, because as much as I could recognise societal inequalities I didn't really experience oppression in my day to day life. I had no conceptual tools to understand my privilege. Clearly, if I wanted to keep feminism in my life, I needed a more complex model and one that wasn't premised on *me*, or my "I", as a centre. (A fourth member of the Feminist Pedagogy Working Group)

Just as feminist critiques from the margins and poststructural thought has shown, women's experience, just like any other experience, is not universal. In order to understand experience, feminists were indeed going to have to look beyond oppressions based on dichotomous notions of gender (feminine/masculine) toward a complexity that values difference and diversity. Feminists have not given up the category of experience, but they have turned toward trying to grasp how concepts related to experience matter in doing feminist research. Thinking about research involves figuring out ways the self can be known, how subjectivities emerge, and how identities form. From these types of understandings, feminists have moved toward identifying, and then learning from, specific subject positionings.

As the writers in this section tell us, taking on feminist research is fraught with irreconcilable dilemmas, unanswered questions, and contradictory practices. Because there is no "one" way or "right" way to think or distinct path to take, a prescription for "good" feminist research does not exist. But at the same time, there are issues that can be addressed that assist in thinking through what it means to take up feminism in research in geography. Drawing on their own experiences in undertaking feminist research projects the authors in this section disclose their feminist research paths, full of decisions they made over the course of their engagements with feminism. Mary Gilmartin tries to make sense of her journey through literature and geography as a way to go beyond the limits of knowledge set for her by geography's colonial past. Meghan Cope details a set of implications arising out of research based on any one of several feminist epistemologies. Louise Johnson, in her work with unemployed women, shows how feminism did make a difference at various points throughout her research project. Together the material in these chapters provides a richly textured blend of a multiplicity of ways to take on feminism in feminist geographic research.

Short 1   Being Feminist in Geography

Feminist Geography in the German-Speaking Academy: History of a Movement

Elisabeth Bäschlin

Scene 1: Pioneering Feminists

In the early seventies, feminist social critics within the women's movement began questioning the prevailing images and roles of women within the family and society as well as the types of research and the dominating discourse in science – first in sociology, then in subjects like linguistics, history, psychology, anthropology and political science. They were pioneers, which means that they had to face solitary fighting, loneliness, and isolation from other women and men (Wagner 1985, pp. 215–25). Feminist geography pioneers in the German-speaking academy were very much the same.

It was only in 1978, in Eva Buff's master's thesis about women's migration out of mountain regions, that for the first time in German-speaking geography, women were treated as a social group separate from men. But it was only in 1982 that a geographical journal, *Geographie heute* (1982), (re)presented women as a distinct social group with special activities and specific fields of action. Typical of the time, the topic was not about women or women's rights in Europe; rather, the focus was on women in Africa, Asia and South America, situations seemingly far away from Europe!

Even more groundbreaking, however, also in 1982, a critical article appeared in a students' journal at the university of Zürich. Anne-Françoise Gilbert and Mechtild Rössler (1982) queried the absence of women in geography. In order to address this absence among a wider audience, the two women organized a student workshop on feminist geography at the official biennial national German conference, known as *Deutscher Geographentag*, or "German Geography Day," in Münster, 1983. Also at the conference, Monika Ostheider, assistant to a well-known professor,[1] gave a lecture entitled "Geographical Women Research – A New Theoretical Issue?" (see Ostheider, 1984). She clearly pointed out the "blindness" in geographical thinking about women's roles in society as well as the lack of women actually either doing geographical research or being the topic of research. Even though located "outside" the official geographical project, women were finally becoming both research subjects and research objects.

After this conference, women students at several universities began engaging with feminist geography. And, after a short time, three feminist pioneer students had submitted the first feminist master's theses – in Giessen, Göttingen, and Zürich (respectively Buschkühl, 1984; Tekülve, 1985; Gilbert, 1985). Even though throughout the German-speaking academy – Berlin, Freiburg, Hamburg, Giessen, Göttingen, Frankfurt, Zürich, Basel, Bern, and Vienna – groups of students and individual women had become active in their departments, there were no feminist geography presentations at the national German conference in Berlin in 1985. Even though several women knew about each other, contacts were still mostly fortuitous; there was still no central organizing group. So, on German Geography Day in München, 1987, students from Frankfurt organized a second workshop on feminist geography. By the end of the workshop, participants, for the first time, made claim on an official "Working Circle on Feminist Geography."

From 1986 to 1988, at the initiative of an active group of women students under the responsibility of the two professors, Elke Tharun and Roswitha Hantschel, a series of lectures on feminist theories of science and women's mobility were held in Frankfurt. Unfortunately, after 1988, due to lack of funding, the lectures had to be given up. In 1989, these lectures were published by the students and became the first book about feminist geography in the German language (Bock et al., 1989)!

Scene 2: Weaving a Network

In spring 1988, Verena Meier, an assistant to a professor in Basel at the time, invited young women geographers that she knew were interested in feminism in Germany, Austria, and Switzerland to come to Les Emibois in the Swiss Jura in order to discuss the situation of women in geography. A dozen women took up the invitation, mostly students, some (the already designated) pioneers. Some of these women had worked for some time on topics in women and geography, but most of them were quite alone and isolated in their departments, missing out on opportunities for stimulating discussions.

In Bern, too, a feminist student group had just been emerging. So, after years of loneliness, I went to the meeting in the company of one of the students. We had hoped to find experienced researchers on women and science espousing a clear feminist research agenda. We had hoped that we could profit from their great experience to build up feminist geography and feminist research in our own department. Oh, how we were disappointed! We found that most of the other women at Les Emibois were seeking the same things: taking great effort to collect information on a theoretical background in feminist geography. We came to realize that were still in the midst of building a feminist geography. Yet, all in all, it was a good feeling to meet other women geographers working from similar perspectives. The meeting gave us an opportunity to exchange stories and to discover how similar our situations actually were.

Most of those active at this time were geography students. None of them had a secure position at a university, or even hopes of getting one. What was happening was that all the young feminist researchers left university at the end of their studies, taking with them their knowledge and experiences, without being able to pass them along to the next generation of students. There was no continuity in research and the production of knowledge. Each student had to start at the beginning; had to "invent the wheel" once more. An impossible situation, but an example of the waste our society makes with women's experiences and women's knowledge.

It was clear that we could not continue like this. We needed women geographers in good positions in universities, as lecturers and professors to build up feminist geography and to assure at the minimum at least some continuity. However, this was beyond our power. Yet we had to do something. So, we decided that we would weave a network to meet our needs. First, we needed a place to announce national and international conferences, publish reports about ongoing research, provide references to literature, and connect women working in the same areas – simply to learn from each other! So we decided to publish a newsletter, the *Geo-Rundbrief* focusing solely on feminist geography. The first issue was published in July 1988, in Bern, because, as lecturer, I was the only one to have access to some institutional support. I published the newsletter until March 2000. In 1998, the newsletter was launched on the internet (www.giub.unibe.ch/femgeo) and has subsequently replaced the paper publication.

Second, we needed some organization to co-ordinate our activities, a group of some kind to bring us together as feminist geographers. On German Geography Day in 1989 in Saarbrücken, we held our first meeting as the "Working Circle on Feminist Geography" with presentations from students doing masters' work (*Diplomarbeiten*). After the conference, we officially announced our existence at the meetings for the German Geographical Association. We did not want our Working Circle to be a strongly organized association; rather, we wanted a sort of "geographical women's movement," held together by the newsletter and our communications. As part of maintaining the network as a group, we considered important our presence at *every* German Geography Day. Since 1989 we have marked the occasion with meetings of our "Working Circle" and, since 1993, with a book display for information as well as a central meeting point. In this way, we tried to establish feminist geography through our concerns about women and our critiques of science and knowledge while, at the same time, trying to serve as a group of contacts to young feminist geographers in order to encourage them to continue their research even in their isolation.

Another important spoke in the feminist network, born at the same time, is the "Student Female Geographers Meeting." The idea for this specific type of meeting arose at a national meeting of student associations in 1989, where there had been no chance to discuss women students' issues. So women students decided to form their own informal, non-institutionalized meeting and met for the first time in June 1989, to discuss their situation as women students in geography. They decided to hold meetings every six months with varying topics about women including feminist geography. All meetings are still being held without any institutional support. At each meeting participants simply decide among themselves who is willing to organize the next one.

A network has indeed been realized. We succeeded in creating a forum about feminist science, in spreading our presence into different geography departments through our newsletter to make clear that feminist geography does exist, and in giving support to students interested in feminist geography.

Scene 3: Entering Institutions

As a small part of "entering" geography, lectures about feminist geography have been held over the past ten years in several geography departments; mostly in response to requests of students and assistants – Frankfurt, Tübingen, Berlin, Vienna, Trier, Klagenfurt, Zürich, Basel. In 1994, the first *Habilitation* in geography with a feminist approach was presented in Basel by Verena Meier. The list of geographical theses and dissertations in feminist geography continues to grow.

We think it is important to bring a feminist approach into the institutions if we want to change research in geography. For us, this means we urgently need feminist geographers as professors. Actually, not only feminist geographers, but simply women geographers, who are also under-represented in universities. Universities in the German-speaking academy continue to be an "ecosystem made by men, for men." Ruth Bördlein's (1994) work supports these observations. She found that most female university teachers in German-speaking universities are over fifty-one years old and single!

Nevertheless, we have begun to enter the institutions! Since April 1997, Doris Wastl-Walter is Professor of Human Geography in Bern. And, in February 1998 Verena Meier became Professor of Regional Geography at the Technical University of München. Both are declared feminists. And this, in spite of all the established professors in Germany who tell young female geographers who wish to pursue positions in the academy simply not to become feminists if they want to have any chance of getting a professorship!

Scene 4: Shaping the Future

As professors, Doris and Verena now have real opportunities to promote feminist geography. They include feminist critiques of scientific discourse in their teaching and develop it in lectures and trainings. Gender research programs can be presented in co-operation with professors from other faculties, such as history and sociology, which gives feminist geography more of a chance to be considered.

We want to work on a feminist reconstruction of geography (Bäschlin and Meier, 1995). In general, our research subjects remain: topics on spatial structures as the relation between access to space and social power from which results dominance or exclusion of social groups (Bühler et al., 1993); questions about the definition of "labor," production and reproduction, and the gendered division of labor, as well as the international one; and topics on social constructions of "nature" and "culture."

In summer 2000, the head of the University of Bern created an Interdisciplinary Center of Gender Studies. The chairwoman of the Center is Doris Wastl-Walter. Great news!

So, feminist geography in German-speaking countries goes its way, becoming more important in the academy.

NOTE

1   In the German-speaking academy, in addition to a doctorate degree, each scholar in a university must complete a *Habilitation*, a major independent research undertaking, or what is loosely equivalent to a second doctoral degree with tenure. In order to pursue an *Habilitation*, a new scholar must work with an established professor, or an "assistant to a professor." Once completed, the *Habilitation* marks the formal entry and acceptance of a scholar into the academy. Because professorships are limited in number, obtaining an *Habilitation* does not guarantee a professorship.

RESEARCH TIP

Data Sources

- Archives
- Case studies
- Census tracts
- Consumer artifacts
- Diaries and journals
- Documents from societies, institutions, organizations, and associations
- Ethnographies
- Focus groups
- Folklore
- Genealogies
- Graffiti
- Icons
- Interviews
- Landscapes (natural and built)
- Life experience
- Music
- Oral histories
- Photographs
- Place names
- Popular media
- Questionnaires
- Surveys
- Video

Last, but not least, be creative and innovative in the types of data you use for your feminist analysis!

2

Making Space for Personal Journeys

Mary Gilmartin

I want to draw a map, so to speak, of a critical geography and use that map to open as much space for discovery, intellectual adventure, and close exploration as did the original charting of the New World – without the mandate for conquest.

Morrison (1992, p. 3)

These are the words of writer Toni Morrison, and she uses them to open a book of essays called *Playing in the Dark*. The essays are broadly concerned with the themes of literature, race and national identity, and Morrison shows the ways in which the language of literature is used to avoid or evade the topic of race in the American context. Once I read this book while an undergraduate in Ireland, it opened a new world for me. As a young girl, I formed my vision of the United States from a strange mix of *The Waltons*, John Steinbeck, and MTV. While I knew about slavery and racism of the past, I had no sense of the ways in which racism had persisted into the present. Morrison's prose and fiction writing showed me this world. Her prose dealt with political issues as diverse as the absence of black people from American history and the furore over Clarence Thomas' appointment to the Supreme Court. Her novels dealt with issues that were specific to black America, such as slavery, the Harlem Renaissance and Civil Rights, but in a way that someone like me – a young white Irish woman – could access and appreciate. Toni Morrison's work opened a space for me to explore all of these topics, and showed me the different ways in which exploration can be used for the purposes of gaining knowledge without appropriation.

When I read this passage for the first time, I was an undergraduate studying Geography and English. In Geography, we took courses in a wide range of topics: from geomorphology to urban geography; from cartography to the history of the discipline. I loved the scope of the subject, but I struggled to find links between it and English. The two subjects seemed to operate in very different arenas, with little interaction. The literature that formed the core of my studies in English – novels, plays and poetry – was discounted as "not quite geography" by many of my lecturers. Morrison's words helped me navigate out of the confusion. We learn through exploration and discovery. We explore by bringing the unfamiliar and the familiar together, in the hope that something exciting will result. And, even though disciplines build barriers and claim territory as their own, we don't have to accept these definitions. We are free to discover; though we often need support as we set out on this journey.

This chapter is the story of making space for my personal journey through geography. Along the way, I have received support as well as discouragement from a wide range of people and places and ideas. I wanted to write about some of these interactions and the ways in which they have influenced my thinking about geography. These influences are not just academic, they are also personal and political. I write as a PhD student, so this journey is far from over. So far, though, it's been both interesting and confusing, and I would like to share that excitement and uncertainty with you. Rather than start at the beginning, though, I want to begin in the present, with the story of a research project that has been running for the last two years.

The Conquests of Geography

I currently work as the research assistant on a project headed by Dr. Priya Rangan of Monash University in Melbourne, and funded by the National Science Foundation. The title of the project is "Common Access Lands and Sustainable Rural Development in South Africa," and we study the economy surrounding the collection and sale of medicinal plants in South Africa. Medicinal plants are used for traditional medicine – the primary source of health care for over 70 percent of black South Africans. Under the system of apartheid, which operated in South Africa from 1948 until the first democratic elections in 1994, people of different races were legally separated from each other. Whites were by far the richest and most powerful racial group, while blacks – the majority racial group – endured poverty and were excluded from political power. White South Africans had access to health care facilities that rivaled those of any wealthy developed country, while black South Africans were provided with rudimentary health care. Given this inequality, black South Africans had to rely on their own resources for treating illness and disease, and they did so by using and building on traditional knowledge about plants.

While traditional medicine is crucial to the health of the country, many people disregard its importance and its efficacy. When we describe our research to people who do not trust traditional medicine, they are often suspicious. They describe traditional medicine as backward, operating in the realm of superstition without any scientific basis. They call traditional doctors "witchdoctors," responsible for barbaric practices such as trade in human body parts and witchburning. They ask, with raised eyebrows and ill-disguised disbelief, "and does it work?" They expect us to answer "no," and to hear about charlatans and bogus medicines. When we tell them that, in our experience, many traditional doctors and their treatments are highly effective, they are surprised and by and large sceptical. They do not want to believe that we have encountered practitioners who have studied for years before qualifying as doctors; who take a holistic approach to the health of their patients; and who operate in conjunction with Western medical practitioners without needing to denigrate their approach. Instead, they want to see traditional medicine as ill-founded superstition, its practitioners as quacks, and its adherents as backward people who know no better.

These reactions are surprisingly familiar to me. As an undergraduate, I became interested in the history of geography. I spent many hours in the library, leafing through dusty old copies of the *Geographical Journal* from the late nineteenth and early twentieth centuries. I encountered detailed accounts of travel, of "faraway" places, and of the people who inhabited them. I imagined the effects of these accounts when they were delivered as lectures, complete with illustrations – stories of primitive people with backward ways of life, who inhabited exotic places full of riches they had not yet learned to exploit. Geographical societies flourished in Europe in this period, as people flocked to hear stories of unknown places. As I was reading accounts of early explorers, I was also reading poetry and prose that dealt with the issue of colonialism for my coursework in English. We read Irish writers such as Yeats and O'Casey and Wilde; we read writers from other colonies such as Fanon and Achebe; we read commentaries on colonialism by writers from Swift to Said. I brought this information to my studies of geography, and as a result started to question not just the practices of geography, but also its history. In particular, I looked again at those early geographers and their active role in colonialism – a process that I had come to view with considerable distaste.

As I explored, I realized that geography had been an academic tool of exploitation. Brian Hudson argued, in 1977, that the project of geography and the project of imperialism were intricately connected, and that geography had prospered because it was useful to the creation of empire. It took many years for Hudson's central thesis to gain widespread acceptance within the discipline of geography. Others noticed the link, though. Edward

Said, in *Orientalism*, provided a broad-sweeping approach to this topic. Drawing on literature and anthropology, history and politics, Said pointed out just how useful geography was in the colonial endeavor. He quoted Lord Curzon, later president of the Royal Geographical Society, who argued in 1912 that geography "is part of the equipment that is necessary for a proper conception of citizenship." Lord Curzon described geography as a handmaid to other disciplines: young, eager, and willing to help in order to prove its worth (in Said, 1978, pp. 215–16). As David Livingstone has shown, one of the ways in which geographers were useful was in telling stories of foreign places. Many stories centered on the inhabitants of these regions, who were portrayed as simple, underdeveloped, or deviant, and in need of salvation. The process of exploration and subsequent colonization was described in terms of a moral imperative to bring enlightenment and truth to these people who otherwise were lost.

By making themselves useful, these early geographers sought to carve out a place for geography within the university system. One such was Halford Mackinder. Within and beyond geography, Mackinder was important in a variety of ways. He worked tirelessly to have geography recognized as a university discipline. His definition of geography as "the science whose main function is to trace the interaction of man in society and so much of his environment as varies locally" (in Livingstone, 1992, p. 190) is still the basis for many school and university courses in the subject. And, as an early practitioner of what we now call geopolitics, Mackinder advanced the Heartland theory. In this, Mackinder argued that control of the Eurasian landmass was vital for world domination. It became obvious to me, studying this theory during a time framed by the Gulf War, that the Heartland theory was still highly influential both within geography and in a broader context. Geography was being used to justify war, and I was sure I didn't like this association.

The space of conquest for the discipline of geography was not confined to its use as a tool of colonialism, or to its desire to become a legitimate academic subject. The writing of geography was equally a struggle over who could be a geographer, and in what way. As an undergraduate in Dublin, studying Mackinder and other great men of geography, I was struck by one particular fact. This has remained in my memory, despite the disappearance of my hastily scrawled reference to its source. Mackinder opposed women's suffrage, on the grounds that women could not expect to have the power of one sex, and the privileges of the other. While that remark had a major impact on me, a young woman struggling to find a place and a voice in academia, I have yet to see it mentioned by historians of geography or geopolitics.

Yi-Fu Tuan wrote in 1976 that "geography provides useful knowledge" (p. 275). In the past, geography was useful to imperialism and colonialism, whether in the cartography of the unknown or in showing the geopolitical importance of specific places. Geography continues to be useful today, through technologies such as GPS (Global Positioning System) and GIS (Geographical Information System). I was concerned with the uses to which this knowledge was being put, and not just in the past. Through my studies of literature, I saw the many ways in which words and language and knowledge could be used to fight oppression. I wanted to imagine a geography that was useful for those without power, and not just for those who wished to justify war or find the right embassy to bomb or exclude women from voting. I believed that geography could be used as a tool for positive change, as a way of writing the world that didn't involve a mandate for conquest.

Anti-colonial Geographies

In my search to understand how geography could be used to resist oppression and to create alternative spaces of belonging, I came across the words of Stuart Hall, a writer of Caribbean descent. Hall wrote that "there is no other history except to take the absences and the silences along with what can be spoken" (1991, p. 48). This passage was echoed by Toni Morrison, who wrote in *Playing in the Dark* that "silence from and about the subject was the order of the day. Some of the silences were broken, and some were maintained by authors who lived with and within the policing narrative. What I am interested in are the strategies for maintaining the silence and the strategies for breaking it" (1992, p. 51). Both Hall and Morrison are writing about race, but their ideas are relevant for a wide range of contexts. I used them to think about an alternative approach to geography – an anti-colonial approach. If colonial geography is about conquest, then anti-colonial geographies are concerned with exposing the reactions to, and effects of conquest. Anti-colonial geographies are concerned with breaking, and writing, the silences of the present as well as the past.

As an undergraduate, I found two different approaches to geography helpful in looking for silences and absences. The first is humanistic geography, and the second, feminist geography. Humanistic geography, emerging in opposition to the scientific imperative of quantitative geography, was concerned with understanding and writing the geography of humans and the humanness of geography. "There must be more to human geography," Anne Buttimer wrote, "than the danse macabre of materially-motivated robots" (1990, p. 5). Instead, Buttimer posited that scholars could be "catalysts for a more mutually respectful exchange among people" (1993, p. 220). An example of one such exchange was a research project that

Buttimer was involved with in Glasgow. There, in the midst of large-scale housing reform, decisions about the mass movement of people to new communities were being made at official and bureaucratic scales. Buttimer and her colleagues talked to people affected about their experiences of forced removal, thus personalizing the planning process and forcing a recognition of the human impacts of planning decisions made by "outsiders." I was lucky enough to have Anne Buttimer as a teacher, and to witness the ways in which she made geography more inclusive and more human, both in her research and in her approach to students. She showed me how geography could be used to benefit those with less power, in a way that involved dialogue rather than an imposition of ideas, by focusing on people's lived experiences.

A focus on lived experiences has been fundamental to the work of many feminist geographers. The concern of feminist geography was – initially, at least – to make geography relevant for understanding the specific needs and experiences of women. Monk and Hanson were early initiators in this project (1982). They accused geography and geographers of gender-blindness. Geography, they argued, reinforced traditional gender roles, avoided research themes that addressed women's experiences, and dismissed the significance of women's activities. Because of the absence of women, both as research subjects and as practitioners, human geography could not legitimately claim to be about humans. Instead, it was about men, though deceptively couched in a language of inclusivity. Feminist geographers have taken up the challenge of breaking silences. They focus on a wide range of topics: from epistemology to research methods; from politically informed research to asserting a place for women within the discipline of geography. In doing so, they have opened the way for a more catholic approach to knowledge, one that is accepting of a wide variety of world-writing.

One area of interest for feminist geographers that I particularly enjoyed was the renewed focus on travel writing. White women travel writers, from Mary Kingsley to Eliza Fraser, were advocated as early geographers. I liked this argument, because I liked the idea of travel writing as geography – it challenged exclusive definitions of what was (or, more regularly, was not) geographic writing. In addition, I was impressed by these women's achievements for they not only endured the discomfort of travel to distant places, but also escaped the restrictions of women's assignation to private places. This new focus on women's travel writing was significant because it addressed a number of silences within geography. First, it insisted on a place for women in the history of the discipline. Women travel writers failed to achieve the same widespread acceptance as their male counterparts, but this does not mean their achievements were any less significant. These women were indeed early geographers, and we cannot understand the story of geography without recognizing their protracted absence. The second silence relates to the types of knowledge that are often not considered "geographic" enough. Women's travel accounts were rarely scientific, and often made no claims to objectivity. As geography struggled to establish itself as a useful science, there was no place for these more subjective accounts of travel by women. By bringing these stories to a broader audience, feminist geographers have opened a space for other, less mainstream geographic voices.

When I wrote my undergraduate dissertation in geography, I was inspired by these examples. I wrote about the travel accounts of an American woman, Asenath Nicholson, who visited Ireland in the 1840s, before and during the Great Famine (Gilmartin, 1999). I like her travel books – they are vibrant and observant and still interesting – even though many (male) critics dismissed her as eccentric and consequently disregarded her work. By focusing on Nicholson's neglected travel accounts, I felt I was playing my part in writing an anti-colonial geography. There is, however, an obvious contradiction in describing early women travel writers as anti-colonial, though I didn't realize it until much later. Many of these women were, either directly or indirectly, part of the colonial project. They could travel in distant places because they were white and relatively wealthy. By virtue of being white, they were associated with the colonizers who, in turn, were among the powerful elite of the places they visited. Their relative wealth gave them the freedom to move around, to hire transport and servants, to eat well and to find accommodation. While their writings certainly provided a different perspective to that of their male contemporaries, I increasingly became uncomfortable with thinking about their work as anti-colonial. I recognize their importance in broadening our sense of geographic writing, but I also see the limitations in relying on these accounts as a challenge to broader inequalities.

It is important for us to question the history of geography, and to ask about the stories we have neglected or the perspectives we have ignored. As I underwent this process, the work of humanistic and feminist geographers was vitally important to me. Humanistic geography stressed the importance of people as thinking, feeling beings, rather than mere numbers or statistics. Feminist geographers insisted on a place for women within the history and practice of geography, and argued for a broader, more inclusive approach to geographic knowledge. I saw both as forms of anti-colonial geographies, because they made previously hidden people and ideas visible again. Yet I also realized that challenges to the conquests of the practice of geography were not sufficient. If we are to practice an anti-colonial geography, we need to look both within *and* beyond the boundaries of the discipline.

Making Place

In an essay by Irishwoman, Nuala Ní Dhomhnaill, I recognized a different form of anti-colonial geography. Ní Dhomhnaill is a poet who writes primarily in the Irish language, and then translates her work into English. "Dinnsheanchas" (1996) is her account of the ways in which Irish people, primarily rural Irish, named and "owned" their surroundings in spite of colonial control. For Ní Dhomhnaill, this happened through *dinnshean-chas*: an intimate knowledge of place. The child of Irish migrants, Ní Dhomhnaill was born in Lancashire but sent, at the age of five, to live with her aunt and uncle in the Kerry Gaeltacht in order that she would be fluent in Irish. Her uncle, Thomas Murphy, fostered her fascination with *din-nsheanchas*. Thomas told Nuala about their family history: where they had lived, when they had moved, what they had worked at – stories that stretched back for seven generations. The crux of Ní Dhomhnaill's argument in this essay is that, through the centuries, *dinnsheanchas* has been a way of knowing the land "emotionally and imaginatively without any particular sense of, or actual need for, titular ownership" (1996, p. 431).

When Ní Dhomhnaill writes of knowing without possessing, she poses an interesting and difficult challenge. Knowledge is territorial. We are encouraged to participate in its struggle for space: to define our areas, our disciplines, our place in the academy. We put up defences. We erect barriers. We fight over what is or is not acceptable as "geography," a battle that is ostensibly about academic rigor, but is ultimately about identity and territory. Clarence Glacken recognized the difficulties with this definitional obsession. In *Traces on the Rhodian Shore*, his account of nature and culture in early Western thought, Clarence Glacken wrote that: "A historian of geographic ideas . . . who stays within the limits of his discipline sips a thin gruel because these ideas almost invariably are derived from broader inquiries. . . . Of necessity they are spread widely over many areas of thought" (in Livingstone, 1992: p. vi).

Glacken certainly achieved this feat. His text is broad-ranging, learned, and inspiring, and takes us on a journey from Aristotle to Malthus and Count Buffon. Through this account, it becomes apparent that Glacken realized the search for knowledge and understanding should not be restricted by the boundaries of the academy. We can learn much about writing the world from those who are tangentially (if indeed at all) related to the academic project of geography. World-writing, though certainly central to geography, is not nor should not be considered the exclusive rights of that discipline.

Feminist geography opened a space for other voices in geography, so that we need not be restricted to a limited way of understanding and representing knowledge. One range of voices is represented by literature. My interest in literature is not motivated by the need to assert just how "geographical" a particular piece of writing is. Rather, I am interested in literature that is open and broad-ranging, and concerned with the intersection between the places we inhabit and the identities we assume. In my conceptual framework, geography is as much about the novels of Toni Morrison, and the film scripts of Roddy Doyle, as it is about superimposing the concentric zone model over land-use maps of Dublin. These writers created, for me, a sense of curiosity about the apparently familiar, a sense of excitement about the unfamiliar. They encouraged me to explore and, in doing so, enhanced my understanding of different places and contexts and relationships.

Written literature certainly adds another, deeper dimension to our understanding of people and places. To rely on written accounts, however, means that we often miss out on the stories and experiences of people who do not express themselves in written form. In academia, we are trained to value the written word over the spoken word, and to validate our research with published references. As a result, we learn to either ignore or discount oral accounts. But for our research in South Africa, much of the detail and richness has come in the form of stories from people who work as plant collectors and traders on a day-to-day basis. We spent time collecting plants with a woman in Eastern Cape who, in a grassland area much smaller than a football pitch, identified over thirty different medicinal plants. We passed hours in the plant markets in Durban, watching and learning about the hundreds of different products that are sold and their myriad of uses. We interviewed plant collectors, plant sellers, herbalists and traditional doctors in five different provinces. These people span a wide spectrum: from urban to rural; from poor to extremely wealthy; from highly educated to having little formal schooling. What they do have in common, however, is a detailed knowledge of their trade. Yet when Priya first suggested a research project that focused on the medicinal plant economy, many people told her she was wasting her time. "It's trivial," they said. "Don't waste your time studying that topic." She persisted, convinced that it was important – for a number of reasons – to understand how this economy works. Her conviction has turned out to be correct. Traditional medicine is extremely important, not just as a source of health care, but also as a source of income for the many people who make a living from this trade. But much of this knowledge is not documented. Instead, it is oral, passed from generation to generation, and through practical experience and learning. And often, this knowledge is discounted: marked as inferior in the same way that women's stories were treated in the past, or just as "natives" were dismissed by colonizers.

This research has taught me the importance of broader inquiries. We have drawn from many different disciplines and from many different people in our quest to understand the economy of traditional medicine in South Africa. Our research is not without problems. We work through translators, so the words of the people we speak to are filtered through others. We are both based at Western universities, which puts us in a privileged position in terms of resources. At times, that privilege can create difficulties, particularly in relation to intellectual property. And we work within a context that is shaped by apartheid, which in turn shapes the ways in which people respond to us and to our research. Despite these problems, however, we have been governed by a curiosity about our research topic, and a willingness to engage with *any* ideas or people that might help us better understand the topic. By being open to new insights, our understanding has been enriched considerably, and we in turn have been able to contribute ideas and information to the people who have helped us. By being appreciative of *dinnsheanchas* in the South African context, we have made a space for discovery that takes place through the stories of others.

A Personal Journey, Continued

> Twenty, thirty years from now, he thought, all sorts of people will claim pivotal, controlling, defining positions in the rights movement. A few would be justified. Most would be frauds ... [What would remain invisible] were the ordinary folk ... Yes, twenty, thirty years from now, those people will be dead or forgotten, although they were the ones who formed the spine on which the televised ones stood. (Morrison 1997, 212)

This passage comes from Toni Morrison's latest novel, *Paradise*. It is a commentary on the Civil Rights Movement, and on the ways in which the history of this or any movement gets rewritten to favor the few at the expense of the many. This process is obvious within the disciplining of geography. The writings of women travellers in the nineteenth century get discounted in favor of the accounts of their male counterparts. Women are written out of "human" geography. The voices of ordinary people are discounted, as the writing of geography takes on a scientific and objective and exclusive tone. Yet, in opposition to these exclusions, other geographers have consistently fought to highlight those who have been ignored in the past.

In my personal journey through geography, I have drawn on these reactions to silences and absences. My engagement with the literature of colonialism and postcolonialism gave me the tools I needed to start questioning geography's imperial past. Literature – both fiction and prose – got me interested in issues of race and identity. The power of writers such as Toni Morrison and Stuart Hall gave me the courage to address these topics in my research, despite being told that it wasn't "geographical" enough.

Feminist geography in particular created space for me, both personally and professionally. Through feminist approaches, I was able to legitimate my interest in literature and race, as well as in the everyday. I realized that geography could just as equally be anti-colonial as colonial; that it could use the language of liberation instead of being bound by the language of discipline and control. These possibilities converged when I worked South Africa, through my participation in a broad-ranging research project that was more concerned with conversation and collaboration than with conquest and control.

I have drawn on many of these interests and concerns in deciding on my own research topic. My PhD is a comparative study of the role of education systems in the crafting of new national identities in Northern Ireland and South Africa. My interest in the concept of national identity stems from studying colonialism and postcolonialism, as well as from an interest in how different groups – blacks in the US, for example – have been excluded from the process of defining a national identity. Education plays a very important role in this process, but often this role is unquestioned. This is particularly true in divided societies such as South Africa and Northern Ireland, where education is seen as one of the tools for positive change. I'm interested in looking at the conflicts that arise over change, especially for groups who feel threatened. In order to understand the conflicts, I speak to people at all levels in the education system, and from very different social groupings. My geography training is useful for this project, but so too is my interest in literature and the everyday and the stories of women and countless other topics that provide different perspectives on this question.

We work within traditions of knowledge, and draw inspiration from what has been done in the past in the name of geography. This past gives us strength and legitimacy. Yet as Claudio Magris writes, "every identity is also a horror, because it owes its existence to tracing a border and rebuffing whatever is on the other side"(1990, p. 38). If we remain within the limits of our discipline, we lose the tools that we need to challenge its horrors and uncover its silences. It is through engagement with other thinkers and writers and critics that we learn to look at what we had previously taken for granted from a different perspective. By writing beyond limits, we learn.

ACKNOWLEDGEMENTS

Thanks to Vincent Del Casino, Carole Gallaher, John Paul Jones, Dervila Layden, Pamela Moss, Nancy O'Donnell, and Priya Rangan for their comments on earlier drafts of this chapter.

RESEARCH TIP

Maintaining a Journal

- Note your ideas as they evolve and your thoughts on different aspects of the research topic.
- Record "ah-has," "oh, yeas," and "of courses" as they come to you.
- Develop a system that works conveniently and thoroughly for your research purposes.
- Organize your journal, e.g. use color coding, index tabs, recipe cards, or different binders to organize your material.
- Subdivide your journal into categories, such as research question, theory, methodology, data collection, analysis.
- Keep more than one journal, e.g. one for methodology, one for theory, one for details.
- Make notes on links among data, theory, and method.
- Enrich your journal by drawings, figures, or clippings. Don't restrict yourself to written text!
- Maintain a reference library (bibliographic software can be useful).
- Keep a detailed record of names, numbers, and addresses.
- Make a note of the commitments you've made to your participants and to the project.
- Date every entry.

3

Feminist Epistemology in Geography

Meghan Cope

What is an "Epistemology"?

Some type of epistemology underlies every research project, yet it is not always (or even often) made clear by the authors of scholarly work. An epistemology is a theory of knowledge with specific reference to the limits and validity of knowledge. More simply put, an epistemology helps us answer the question "How do I know what is true?" (McDowell and Sharp, 1999, p. 75). For example, if we decide only to validate knowledge based on direct observation, we have identified an epistemology: we can say "we know X is true because we can observe it." This particular epistemology (called empiricism) is fairly simple, yet already we run into problems in many areas of research. For instance, it may be difficult to observe phenomena such as racism, the global economy, or democracy, yet we know they exist because of their *effects* on society, economics, and politics.

In order to understand what epistemology is and how it influences research, we first need to recognize that knowledge is humanly constructed or "produced." That is, knowledge is not just "out there" waiting to be revealed to us; rather, we are active participants in producing what counts as knowledge. Once we understand this point, we can move to recognizing that if knowledge is indeed produced by human actors, there must be multiple and even contradictory perspectives, interpretations, and uses of knowledge. Knowledge is not something that we can passively or actively *acquire* because we are always involved in its production and interpretation. Similarly, knowledge production is never a "value-free" or unbiased process.

A third point arising out of the two premises above is that there are many epistemologies possible in the research process. Researchers' individual perspectives will inevitably influence their *privileging* of different types of knowledge (for example, a researcher will value some sources, forms, or kinds of data more than others). Privileging certain types of knowledge shapes the researcher's understanding of the results and in turn influences the way that others will understand the published interpretations of results.

There are many levels at which we can identify ways that the underlying epistemology biases and shapes research processes and results. Researchers make decisions about how to form their research questions, how to collect data, how to analyse data, and how to communicate results to a wider audience. At every point in this process, the beliefs and interests of the researcher matter to the outcome. In the past few decades there has been a push to reveal and make explicit the underlying epistemologies of all areas of research because of the growing recognition that all research is rife with biases that researchers bring with them to the research process: culture, race, gender, class, and other forms of difference. While we recognize that there is no such thing as bias-free research, there is a growing commitment to minimize harmful biases of omission or discrimination (e.g. ignoring the influences of race or gender on disease statistics) and to freely acknowledge other biases (e.g. coming from a middle-class background in a study on poor families).

So What is a Feminist Epistemology?

If we understand an epistemology to be a theory of knowledge that is particularly concerned with how knowledge is produced and its inherent biases, we can identify epistemologies coming from various perspectives such as an anti-racist epistemology, a Marxist epistemology, or – as is explored here – a feminist epistemology. There are two aspects to a feminist epistemology according to Elizabeth Anderson (1995). The first involves how the consideration of gender influences what "counts" as knowledge, how knowledge is legitimized, and how knowledge is reproduced and represented to others. For example, the recent push in feminist scholarship to "listen to women's voices" challenges the ways that previous researchers discounted women's words as qualifying as "knowledge." To take this notion a step further, a feminist epistemology involves not only hearing "women's voices" but also thinking about how gender as a set of social relations affects both men's and women's responses in the research frame-work, how the genders of interviewers and respondents might affect the data, and how research results are circulated to academic audiences and the public.

The second aspect of feminist epistemology requires thinking about how socially constructed gender roles, norms, and relations influence the production of knowledge. What do we as a society consider to be legitimate knowledge? And what role does gender play in the production of knowledge? How is our understanding of knowledge production influenced if we start thinking of multiple genders along a spectrum rather than as a binary of women/men? If the production of knowledge is an active process involving differently situated human actors, we would expect that people's various experiences, identities, and social locations will influence what they count as knowledge and how they participate in its production and legitimization.

Gender influences the ways that people experience the world, interact with others, and what opportunities or privileges are open or closed to them. One of the most important elements of gender relations is the way that they solidify hierarchies and relationships of *power* in a society, through various means of, on one hand, *oppression* (violence, discrimination, marginalization) and, on the other, *privilege* (preferences, favors, power over others). Therefore, in the processes of producing knowledge, we would expect those who are oppressed to have different roles in constructing and legitimating knowledge than those who are privileged and in power. Most likely, the views and thoughts of the former group will be subsumed under those of the latter group. This is an example of considering how gender relations influence the production of knowledge: women's active participation in what "counts" as knowledge has historically been seen as less significant than men's through the mechanisms of power-based gender relations. That is not to say that women are less active in or capable of producing knowledge; rather their roles in this process have been unrecognized or discounted due to the exaltation of men's roles.

A gendered analysis of knowledge production is more complicated than merely looking at the roles of men and the roles of women in making knowledge. Gender affects societies deeply and in multiple ways that are not always easily identified, separated, or categorized. Gender as a set of relationships influences the production of knowledge through many avenues: media, the socialization of children, religious and cultural values, and political and economic processes. Part of the task of understanding the influence of gender on the production of knowledge is to try to identify and tease apart these many influences while remaining conscious of the ways that these are constantly changing and affecting each other. Further, by opening our inquiries to include multiple genders and not just those of women and men, we introduce another level of complexity to a critical approach to the ways that gender relations influence the production of knowledge. Overall, a feminist epistemology takes gender as central to understanding the production of knowledge and thus influences the nature of research performed and interpreted from this perspective.

Challenging Masculinist Science

The first task of understanding and adopting a feminist epistemology is to reveal the *masculinist* underpinnings of science, particularly the ways that claims to "value-free" or "neutral" science actually mask gender significance. Gillian Rose (1993, p. 4) identifies masculinist work as that which claims "to be exhaustive and it therefore thinks that no one else can add to its knowledge."

Consider a labor market analysis that looks carefully at all the obvious variables of where different jobs are located, what the qualifications of people holding those jobs are, journey-to-work times, and the increase or decrease of job availability over several years. This analysis may be thorough in some people's eyes, but a feminist researcher would immediately want to know how gender affects the workings of the job market, asking such questions as: Are women concentrated in certain jobs and excluded from others? Do women and men earn the same amount for comparable work? Do women's greater household responsibilities affect their labor market participation? Are women's journeys to work shorter or longer than men's? Is there a gender division of labor? And, more broadly, how do gender relations saturate the practices of the labor market in countless subtle ways, ideologically, culturally, and politically? By ignoring these questions (and others, such as what effects race and racism have on the labor market), the initial study seems inherently incomplete and flawed because it was based on the assumptions that not only do individual women and men behave the same way in labor markets, but also broad gendered social relations have no impact on employment. This is an example of the limited vision of masculinist science. Although the researchers undoubtedly felt they were being "neutral," objective, and thorough in their research, their failure to consider a wide range of questions and issues relating to gender renders their analysis partial and narrow. A feminist critique of this research challenges its epistemology as masculinist because by failing to examine differences between men and women and the impact of gender relations on work and labor markets, the authors assumed that all practical and ideological impacts on labor markets had been fully accounted for in the analysis.

Another aspect of masculinist science is an assumption that men comprise the norm for humanity. The pharmaceutical trials of aspirin in the US demonstrate this point clearly. Pharmaceutical trials for new drugs and old drugs for new use are considered to be some of the most rigorous and "bias free" in all of science. The US Food and Drug Administration is very careful about allowing drug companies to make claims about the effectiveness of their products based on their own research. However, in the case of the use of aspirin for preventing subsequent heart attacks (the results were widely circulated via television and magazine advertisements), the initial study was done entirely on men. The researchers failed to consider the application of this treatment for women and yet the results were published as if they applied equally to everyone.

So what do we mean by "science" here? Science is perceived as systematic, unbiased, neutral, rigorous, and ultimately, the best way to get to *the truth*. Yet from the examples above, we know that these perceptions of science are open to question. Many of the roots of modern Western science lie in the European Renaissance era when philosophers and researchers attempted to use empiricism (direct observation) to throw off the veil of mysticism prevalent prior to the Renaissance. The notion of truth developed such that it was assumed to be pre-existing (out there for us to discover) if only we could gather enough evidence and measure phenomena using enough variables. However, feminists and postmodern scholars have critiqued the empiricist approach of "science," claiming it produces merely one of many competing "truths." The very term "science" has had a long history of masculinism because it has represented a powerful force in society that has consistently ignored or actively suppressed diverse forms of knowledge production, the importance of gender and other sets of relationships on constructing multiple truths, and, finally, "science" has carried with it the assumption that its complete and exhaustive authority over knowledge cannot be challenged.

The Science Question

Sandra Harding (1991) developed three critiques of masculinist science. In one sense, there is the matter of simply doing "bad science" by creating and maintaining highly discriminatory hierarchies within science disciplines that prevent women (and other people who are perceived as "unscientific") from reaching the upper echelons of performing research. Second, Harding considers "standpoint theory" in which there is an understanding that multiple perspectives are valid in as much as they are genuinely held by people coming from different "standpoints." From this perspective, "science" can be critiqued on the basis of the limitations of male researchers' understandings due to their positions and experiences of power and privilege. That is, male researchers cannot possibly understand multiple views of a problem because the world is at their feet and there is no need to step down from a pedestal of power (or so the theory goes). In a third critique of masculinist science, Harding reviews the contributions of feminist postmodernists who challenge the very premises of science as merely expressions of power and oppression. That is, the whole project of science is just a mechanism to keep large numbers of people under the thumb of a few elites. Of course, these three critiques are overlapping rather than mutually exclusive – standpoint theorists may well agree that science is a mechanism of the elite, for example. The important point is that the concept of science is challenged on the basis of its masculinism through many different routes, both practical and ideological.

But where do we go from here? Is science salvageable in any form? Or must we all just wallow in a sea of subjectivity and relativism with no universal truths upheld by the public? Harding and others argue that the key to saving science is to recognize that objectivity cannot be increased by some pretended "value neutrality." Rather, we need wide and rigorous inclusion of multiple perspectives, including those of the oppressed and marginalized. She and others have advocated a "strong objectivity" that maintains the goals of rigorous scientific inquiry but requires a wider array of questions, interpretations, different perspectives, and inclusion of researchers and subjects from marginalized groups to strengthen the claims of "truth."

As another response in challenge to claims of value neutrality, Donna Haraway uses the concept of "situated knowledges" (1988). By this, Haraway means that we must reject the all-encompassing "truth" notion in favor of context-specific and situation-sensitive knowledges. For instance, rather than searching for universal statements that apply everywhere to everyone (and therefore really apply nowhere and to no one), it would be better for us to acknowledge the biases, perspectives, and contextual factors such as political systems and cultural values inherent in the research project and move forward from that point. Harding, Haraway, and others are attempting to demonstrate that the cloak of "neutrality" in science is actually hiding the true complexities of research in our real world of messy and complicated phenomena and masking the perpetually and necessarily partial views that we as researchers hold.

Feminist Epistemology in Practice

How does a feminist epistemology make a difference in a research project? First, we need to acknowledge that there are many possible feminist epistemologies; that is, there is no one right way to do feminist research. Just as there are multiple forms of feminism, there are multiple feminist epistemologies that may, even in a given project, complement, contradict, or build on each other. In a general sense, any research project involves forming research questions, collecting data, choosing methods, analysis and interpretation of data, and representing the results – in each of these stages a feminist epistemology *matters*.

Forming research questions

Before a feminist researcher can even go about collecting data, she or he must formulate the research question(s) that guide the project. For example, a general interest in how people use and experience a public park would depend on various factors including the ages, race, class, gender, sexuality, ability, etc. of the users, and on the purpose of being in the park (to play, to sleep, to meet people, as a short-cut to work, to buy drugs). These issues may then spur the feminist researcher to recognize that relations of power, such as gender, make a difference in how/when the park is used. Following this, the researcher may consider the broader question of how construction of gender norms, expectations, and relations influence perceptions of public space. This is how epistemology helps shape a research question – by raising issues of the significance of gender right from the start.

Collecting data

Gathering data for a research project involves many steps as well, which, again, are influenced by epistemology. First, the feminist researcher needs to consider what kind of data would be appropriate to connect to her/his theoretical framework. For example, someone doing a study of housing issues for Hmong refugees in California would need to consider people's own preferences, forms of discrimination in the housing market, loan availability, job locations, transportation, cost of housing, citizenship issues, language barriers, etc. The researcher would then need to determine exactly how to measure or create indicators for each of these categories and figure out where the data would come from – the census, a mail survey, in-person interviews, participant observation, bank data, and so on. Epistemological questions in data gathering come through the consideration of what "counts" as data. In feminist research, data issues are influenced by an explicit emphasis on legitimizing women's knowledge, exploring gender as a set of power relations, valuing gender as a central variable in quantitative studies, and considering ways in which social constructions of gender influence the production of knowledge (*Professional Geographer*, 1994). Indeed, in many ways feminist scholars have pushed the boundaries of what "counts" as data, using diverse sources such as diaries, letters, photographs, songs, and artwork to broaden our understanding of women's lives and gender relations, particularly when few other sources are available for capturing their voices. For example, these have been especially helpful in historical work because women's history is less often recorded, and rarely in their own words because masculinist history/

geography has been *privileged* in the production of knowledge, in part by privileging certain data sources over others (see Norwood and Monk, 1987; Katz and Monk, 1993; Domosh, 1996; Cope, 1998).

The actual acts of collecting data are also implicated in the epistemological basis of feminist research. For example, who is interviewed, what questions are asked and how they are phrased, the gender of the interviewer *vis-à-vis* the gender of the respondent, the nature of the interview (formal list of questions versus an open-ended conversation), and even where the interview takes place (a formal office, a coffee shop, a public space, a private home) are important elements to the construction of the research project. These issues all have implications for epistemology because, again, choices about performing research in certain ways and not others indicate what we consider to be the limits and validity of knowledge. These factors are also important to acknowledge in the representation of results because they have a great deal of influence on the findings. For instance, Moss (1995b) showed that her gender facilitated research with the house cleaners she interviewed and observed (though her class status complicated it somewhat), while England (see chapter 12, this volume) found that her gender hindered her interviews with male banking CEOs in a highly patriarchal setting.

In terms of data, the feminist imperative to challenge "masculinist science" requires entertaining a wide array of explanations for one's hypothesis or theory, and then thinking about how the data collection strategies reflect that commitment to feminist inquiry. Attention must be paid to all stages of data "mining," collection, and validation.

Choosing methods

A feminist epistemology does not require the use of *specific* methods, but it does require critical reflection on the use of all methods of analysis and interpretation. The combination of a set of methods with a particular epistemology is commonly referred to as the "methodology" of a project. As with any research, the methods should be appropriate for the data and for the research questions. Interviews and personal logs may be more appropriate for learning about women's experiences of gender discrimination in employment while large-scale data bases and statistical analyses are more suitable for demonstrating the broad effects of labor market biases (Lawson, 1995). However, both qualitative and quantitative methods can be imbued with a feminist epistemology that shapes the research questions, sees gender as an important set of relations that has deep repercussions on the lives of both men and women, and considers the ways that gender influences the production of knowledge.

Similarly, the methods used for data analysis should be appropriate for the purposes of the research. If the purpose of the research is to understand the social and economic implications of women as subsistence farmers then certain methods may be more appropriate than others – for example if "official" data sources show very few women engaging in farming but casual observation suggests that the practice is in fact very common, then dependence on large, government information and statistical methods would not be adequate. And this data mismatch might suggest a good place to start a research project!

Analysis and interpretation

Analysis refers to more than merely the methods used to analyse the data, whether statistical regression or qualitative text analysis. There is also a component of interpretation, reflection, and re-evaluation that involves thinking about the meanings and implications of the data rather than merely the results. First, a feminist analysis is sensitive to gender differences in social, political, and economic relations for both women and men. For example, a feminist epistemology requires that we acknowledge that men and women navigate labor markets, transit systems, health care, farming practices, and cultural traditions differently and leads us to seek to understand the root of these differences and their impacts on daily life and beyond.

Second, feminist researchers are sensitive to using gender as a "problematic"; that is, seeing gender as the central hinge of the research on which everything else pivots. This level of analysis involves considering differences as part of a larger system of gender relations that are deeply embedded in social, cultural, political, and economic processes and maintained through everyday practices, beliefs, and expectations, as well as structural forces such as laws and institutions.

Finally, analysis in feminist research is sensitive to how a gender perspective influences the production of knowledge. Here, analysis becomes a highly reflective process in which a researcher acknowledges her or his own gendered perspective and how that shapes the interpretation of results. This is akin to Haraway's concept of a "situated knowledge" where the *context* of the researcher, the subjects, and the place (both social and physical) are taken into account in the analysis to understand how gender influences the production of knowledge: who produces "legitimate" knowledge, how it is produced, whether and how that production is contested, and in what broader context knowledge is created and re-created.

Representation of results

The communication of the results of data gathering, methods, and analysis
to a broader audience is not a simple task free of biases. Whether it is in a
written paper, an oral presentation, or through some other form of media
(film, website, policy report, etc.), the representation of results is complex
and epistemologically significant. Feminists have long challenged the notion
of "experts" who spread "the truth" in all realms of science; such experts
are critiqued as validating only some forms of knowledge, masking or
ignoring their biases, and using the power of their positions to avoid
opposition in the construction of their truths. In response, feminist
researchers have developed various strategies that have been used to disrupt
the traditional power dynamic between researcher and subjects, such as
sharing and confirming the results of analysis with the respondents and
even co-authorship of the final presentation. These strategies are not free
from problems either, but they are worth considering for certain types of
research. A key point is that the representation of results should be
approached with as much care and self-critique as any other step in the
research project. Feminist epistemologies make a difference to the representation stage; in effect, the representation of research is producing knowledge and is therefore highly influenced by the underlying epistemological
concerns of the author(s).

However, even here we need to disrupt our tidy notions of what
constitutes "representing results." Feminist researchers have been on the
forefront of exploring alternative research strategies, which influence all
stages of the research process including representation of results. For
example, co-authoring with subjects, participatory action research, and
other forms of collaboration destabilize the traditional model of researcher–
researched and create openings for new forms of communicating findings.

Finally, the media through which the data are shared and the audience
to whom they are targeted are important considerations. While much of
academic research is published solely in disciplinary journals (often for the
benefit only of the researcher who gets a degree or a promotion based on
publications!) or presented at academic meetings, there are other forums
through which research results can be effectively spread – and perhaps
make a greater difference in people's lives – such as press releases to
newspapers, television news spots highlighting special events, reports to
political decision-makers, and internet sites. The process of communicating
research findings is always politically and epistemologically complicated.
As Pamela Moss has written, "we [as researchers] must be able to come to
terms with *differences in the way power is constituted* ... between the
researcher and the 'researched'. . . . *Difference* embodies oppressive and

exploitative relations and the experiences of the processes of marginaliza-tion" (1995b, p. 83; emphasis in original). That is, just because feminist researchers try to be inclusive in the writing or production process and try to engage at a political level with empowering their "subjects" does not mean that power and oppression are eliminated. While various researchers approach this problem differently, common strategies include maintaining a self-critical reflexivity (questioning one's own actions and motivations in the research) and using existing privileges of the researcher as ways to foster change (e.g. using the "authority" of a university setting to call attention to oppression, abuses, and exploitation). The work of a critical geographer is never done!

Feminist epistemologies in geography

Feminist geographers have, in the past twenty years, dealt with all of the issues discussed above both by actually doing research and writing about it, and by reflecting on the research process and writing about that. In terms of actually *doing* geography from the position of a feminist episte-mology, I would like to provide two examples by reviewing two quite different projects that use gender as a central problematic.

The first is Susan Hanson and Geraldine Pratt's multi-year project designed to examine the connections between work and home for women and men in Worcester, Massachusetts, in the late 1980s and early 1990s (Hanson and Pratt, 1995). In this project, gender is seen as a central dynamic through which the social and the spatial are mutually constructed. They argue that the gendered practices of the household division of labor and the labor market's occupational segmentation are inherently dependent on each other – women's socially constructed duties in the home serve to put both spatial and time limits on their job-search activities, and, simul-taneously, cultural gender expectations mean that employers are more likely to hire women for particular jobs and under particular conditions (e.g. clerical work at part-time for low pay and no benefits). In this case, a feminist epistemology *made a difference* to the research in so many interlocking ways that it is nearly impossible to conceptualize what their study would have been without it. Certainly we would not have such an elegant exploration of the deeply embedded nature of gender in labor markets, nor such a thorough demonstration that the links between home and work are extremely important for constructing both women's and men's lives. From the first moment of conceptualizing a research question ("How are gender differences constructed spatially?") to the final writing ("Our argument is that social and economic geographies are the media through which the segregation of large numbers of women into poorly paid jobs is produced and reproduced" (Hanson and Pratt, 1995, p. 1)), the authors highlighted women's and men's different experiences, used gender as a central set of relations and constructions, and considered the ways that gender matters for the production of knowledge.

The second project involves a grassroots political organization, the Kensington Welfare Rights Union (KWRU), that is concerned with economic human rights violations, particularly their impacts on poor women and their families in the Kensington area of Philadelphia. This project is currently being conducted by Melissa Gilbert and Michele Masucci at Temple University (Gilbert and Masucci, 1999). The researchers classify their project as "action research" to acknowledge its significant component of political activism. The project's feminist epistemology is demonstrated not only through the centering of how gender roles, relations, and identities influence the situations of poor people, but also through the positions of the researchers and their students as facilitators of political change. Therefore, all stages of the research process are intertwined with activism, teaching, the functions of the KWRU, and empowerment of poor people. For instance, Gilbert brings materials about the KWRU to professional talks, Gilbert and Masucci co-teach a "service learning" course in which students get experience with and simultaneously learn about community organizing and uneven development in cities, the researchers' students record poor people's testimonials to document abuses and conditions, Gilbert and Masucci have used their expertise and access to technology to set up an information management system to catalogue these testimonials, an internal website was developed to display data for other poor people to use, and several workshops have been run by KWRU with assistance from the researchers and their students (Gilbert, personal communication, November, 2000).

In this project, it is virtually impossible to separate out the many ways that a feminist epistemology has saturated the research or even to identify clearly what the "representation of results" means. By sharing the process of knowledge production between researchers, students, KWRU members, and poor people themselves in a fluid and recursive manner, Gilbert and Masucci demonstrate that there is a great potential for the rich nuances of diverse experiences to be explored and represented, not by a single, privileged authorial voice, but through a more collective effort resulting in reflective narratives, policy directives, theory-building, and political empowerment.

Beyond Gender to Multiple Forms of Oppression

Feminists concerned primarily with gender quickly discovered that it was very difficult to talk about the oppression of women without also making reference to racism, heterosexism, and oppression based on disability, religion, age, culture, class and other forms of difference. Fighting oppression on one axis makes little sense when there are multiple forces at work and the effects of each are impossible to separate. Is a particular woman living in a poor neighborhood and working at a low-wage job because she is a woman? Because she is Latina? Because she and her family have always been poor? Because she has a disability that limits her job and housing options? These things are impossible to separate out in a single woman's life because they are so deeply intertwined in her identity and her experiences of these things are inseparable due to multiple forms of oppression. Consider your own identity – are you able to separate your gender from your race, from your class, from your religion, from your sexuality, from your age? Or are they all mixed up together in your *self*? This is the realization that feminists have come to (slowly) and has forced us to think not only of how to challenge gender oppression, both politically and academically, but also how sexism is just one aspect of a kaleidoscope of oppressions that require a united strategy of analysis and critique.

Anderson (1995) suggests that we must extend our view from just gender to include the many and varied forms of oppression that are socially and spatially constructed. This extension forces us to ask research questions differently ("How are gender *and* race twined together in people's experiences?"), suggests a re-evaluation of our data collection procedures and the methods we use for analysis and interpretation, demands that we represent our research in ways that are sensitive to all the multifarious forms of oppression that influence the processes, people, and events we study, and, ultimately, makes us reconsider the ways that knowledge and truth are socially and spatially constructed, produced, and critiqued.

ACKNOWLEDGEMENTS

The author thanks Pamela Moss for insightful comments, patience, and support during the writing of this chapter. Thanks too to Melissa Gilbert for giving me a window into her and Michele Masucci's Kensington project, and to anonymous reviewers for helpful comments. In memory of Sara and Emily, who would have been very good feminists.

RESEARCH TIP

Gaining Access to Research Participants

- Make a cold call.
- Use your contacts (networking).
- Obtain documentation outlining institutional support for your project.
- Prepare a brief description of your research project for quick distribution.
- Be prepared to substantiate your credentials.
- Follow up on initial forays into making research contacts, i.e. don't leave someone hanging and don't make undeliverable promises.
- Produce information requested by interviewee promptly.
- Customize contact letters.
- Demonstrate the value of potential interviewees' participation in your project.
- Gather appropriate information to demonstrate awareness of the fit of the potential participant in the research project.
- Be persistent.

4

The Difference Feminism Makes: Researching Unemployed Women in an Australian Region

Louise C. Johnson

I am a geographer and a feminist who has been doing feminist geography for more than ten years. This has involved documenting the patriarchal economy of Australia's textile industry (Johnson, 1990), reading the gendered spaces of suburban houses and shopping centers (Johnson, 1993, 2000) and uncovering the politics of difference on Melbourne's urban fringe (Johnson, 1994a, 1994b). Each of these feminist and geographical projects has been structured by a further theoretical perspective: socialist, postmodernist or postcolonial feminism. The research in each case – the questions asked, the methods adopted, the conclusions reached and the style of presentation – all varied as a result, but their feminism remained constant. As such, each research project was underpinned by some passionately held assumptions: that gender matters, that women are both different from men and oppressed by them, and there is an obligation not only to describe this situation but also to change it. Doing feminist geography research for me therefore has involved a set of clear and prescriptive guidelines that have no doubt colored the research process. The foregrounding of such assumptions along with the active place of the researcher within the process are two of the great strengths of a feminist approach. However, as a result of assessing feminist research grants and by reflecting on my own recent work, I am increasingly convinced that such assumptions – or subject positions – present profound limitations. Assessing feminist research applications has led to a spirited questioning by others – usually non-feminists – of the predetermined nature of this research such that there appears to be no real testing of patriarchy if it is already assumed. And it has proven difficult to dismiss such arguments, raising real questions about the rigor and veracity of such feminist research. Reflections on my current research into women's unemployment have raised similar questions on the politics of doing feminist research – the power relations involved and the assumptions made. These questions have converged on the process of recruiting and working with a female research assistant and in analyzing interviews with a group of unemployed women whose age, class, disability and ethnicity appeared more significant than their gender in shaping their experiences. The results thus far of such reflections are presented below.

The Project: Researching Unemployed Women in an Australian Region

For some years now I have been researching the place of women in regional labor markets, focused on the city of Geelong – a center of 160,000 people 70 kilometers west of Melbourne in Victoria, Australia. Most recently I have located this work within a postmodern framework that details the role of ethnicity, age, physical and mental capacity, education and geographical location in shaping women's entry into the labor force and employment-related identity. The project began by correlating national to regional trends in employment change. This overview confirmed the movement of the city's economy from manufacturing – car and truck building and the making of textiles, clothing and footwear – to one based on business, community, personal, recreational and tourism services. Such an excursion into census tabulations was not particularly feminist beyond its focus on the gendered patterns of these employment changes (a feminist empiricism); though it was geographical in its concern with the regional specificities of national trends. Having set the general economic context, the project's main aim was to record through face-to-face interviews the experience of a differentiated sample of women as they attempted to enter or re-enter the expanding service sector. Here the objective was to document employment histories, record women's transition from unpaid to paid work and uncover the many aspects of identity mobilized in such a move for mature aged, young, non-English-speaking and disabled women. The academic context was literature on regional restructuring and feminist debates on identity formation and difference. A paid research assistant conducted the main work of interviewing women.

The Relationship: Positioning the Researchers

How the research was conducted is related to the changing nature of enquiry within the academy. An increasing imperative within the cash-starved Australian university system is the measurement of research prowess not by publications, editorial activity, supervision of research students, contribution to public debate or professional conferences, but by winning

competitive funding. Research projects are therefore driven less by intellectual curiosity than by the need to assemble a team that can secure a hefty
budget for travel, equipment, and paid staff, preferably from outside the
university. Not only does this system change the way research is conceptualized but it also builds into it a labor relation between employer and
employee which for many women researchers is a major challenge. For
myself as socialist and feminist from a working-class background, this shift
produced a number of dilemmas. It took some years to rethink the research
I had been doing into a form that required funding but even longer to come
to terms with the politics of employing someone to *do* the research. And
this was primarily an issue of class – in that as a socialist I saw the
employee–employer relationship as essentially an exploitative one. The idea
of entering into an employment contract as well as a collegial research
relationship was oxymoronic and a source of ongoing moral tension. It
meant problems in recruiting people – in that I didn't really want to do it,
saw it as a way to help "needy" others (with money, training or experience)
– and in giving direction – as I was often guilt stricken over the idea of
anyone doing the research I should be doing. And in this there was also a
sense of loss and of alienation from the research process. For now someone
else was taking over a highly personal agenda and in the process inevitably
compromising a style that was highly intuitive, pleasure driven and one
which had benefited as much from random events as from systematic
enquiry. Employing someone to do my research work therefore appeared
to inevitably create an exploitative labor relation, produce research that
was not my own, that was not as "good" as I could do and robbed me of
the pleasures as well as the agonies of its doing.

These anxieties led to ongoing ambivalence in the relationships I developed with a number of research assistants. They led to selecting the
"wrong" people to employ – driven by needy cases and friendships rather
than by rigorous appraisals of who was experienced in giving research
assistance – and to employees who floundered as a result of poor direction,
non-existent timelines and unclear outcomes. It also produced large
amounts of data which have never been analyzed – good in quality but
created at a distance, without the immediacy or urgency of a project driven
by personal passion which develops its own drive for analysis and a closure.
There was also the pressure to move from one budget year to the next,
from one project to another, so that the time needed to amass information,
sift it, think about it, and write about it does not exist, swallowed as it is
by the need to do the next grant application and project. The financial
imperatives driving much research work in Australian universities have,
therefore, produced a whole series of dilemmas. It has taken me some time
to learn how to be a manager of research and of people – to be an employer
as well as an academic. Who to employ and under what conditions remains

an ongoing issue for me as a socialist feminist researcher, though the labor relation aspect is far easier as a result of now heading a department with forty staff!

Having successfully dealt with the moral dilemmas associated with actually employing someone and directing them into a viable research project, the issue for the work on unemployed women in Geelong was recruitment and the nature of the relationship. As a feminist whose research focus was on identity issues and the experiences of unemployed women, I saw it as obvious and desirable to employ a woman with relevant experience, feminist politics and connections into the group to be interviewed. I had worked in Geelong first on my PhD in the 1980s, examining the fate of women and men in the city's textile industry (Johnson, 1990) and had subsequently done research on social polarization in the city and on its shift from a manufacturing to a service economy (Johnson, 1994a,b, 1996). I also had worked with a community employment service provider writing a pamphlet arguing for the importance of the regional perspective in understanding Australia's unemployment problem (Johnson, 1997). It was from this latter connection that I came in contact with Catherine Newton – the unemployed wife of one of the workers who had commissioned the pamphlet – who effectively became the partner investigator in this project. Availability and personal contacts therefore led to a mature aged woman; a trained psychologist who had worked as a financial counselor within a number of community employment agencies for unemployed women in Geelong. Her age, professional training and links into the relevant community networks meant that there was a real professional equality between myself and Cathie, the research assistant. But these same characteristics meant there were class and other differences – of knowledge, professional standing, and association with employers and charitable agencies – between Cathie and those being interviewed. Past experience with the employment agencies as a counselor meant that she had occupied a position of some power over but had also supported women using the agencies. Further, her educational and professional background meant a real distance existed between respondents and interviewer. However, through her sensitivity, social commitment and counseling experience, these social barriers were actively negotiated and assisted in creating an atmosphere in which personal information could be shared. We would meet regularly to talk about these issues and Cathie made it routine to write down the "methodological dilemmas" she encountered. Awareness and articulation of such standpoints is a fundamental and invaluable requirement of any feminist research project; positioning the researchers as well as their "data." The exact impact of these positions on the data generated though is far harder to determine. Did it lead to more profound questioning or more guarded answers? I have no way of knowing for this project. However, the importance of the sex, professional training, class position and outlook of any research assistant compared to another working on the same question would be a worthy research project in itself; to detail the ways in which the data varied (or did not) as a consequence.

The choice of such a research assistant raised a number of issues as the project unfolded. As we came from a similar class and professional background, had comparable commitments to feminism and to the worthiness of the project, it became a relationship that was more of an equitable partnership than an employer–employee relation. This resulted in a sense of co-operation and equality of skill and expertise that made meetings easy and honest debate possible. Indeed at some point in the future – when both of our work lives are simpler – we are committed to co-authoring papers from the material generated. However, there were also down-sides to this relationship as it inevitably remained one of employer and employee – so that each meeting would involve personal gossip, talk of the project and a sharing of thoughts; but also the reporting of work done and the signing of time sheets by her for me. Even more seriously, the initial grant application of A$8,300 had been trimmed to A$5,300. As the research grant was primarily to employ a research assistant to conduct interviews and process the results, the reduction in funds meant that a half-time job had to be reduced to one day per week and remunerated at an hourly rate rather than as a salary. This led to far more administration and insufficient paid time to complete the task. The outcome was a reduction in the project scope – a cut in the number of interviews from 50 to 10 – and supplementary funding to remunerate a research assistant who became deeply committed to the project and willing to work for its successful completion rather than money! Timing of the project also meant that monies had to be spent in a particular calendar year, yet the work extended into the next. To deal with this, an advance payment for work occurred; but it is doubtful if all the work that was subsequently done was truly paid for. This outcome was the result of the partnership and caring for the feminist cause that was part and parcel of our relationship and the project itself. However it was also bad employment and feminist practice and offers a cautionary tale for others.

A further issue is who then "owns" the data generated? We collectively generated the two question schedules – one to collect some basic demographic data and a longer set of interview questions – but then Cathie did all of the preliminary questionnaires, the longer interviews and then transcribed the interviews. It was only at this point that I re-entered the project, to "analyze" the questionnaires and the interviews. In this process and in writing up the "results," I have asserted ownership over the information and effectively claimed it as my own. The tendency to obliter-

ate the labor that generated it is understandable but inappropriate in any
research – feminist or otherwise.

The Interviews: Positioning the Data

The interview phase of the project proceeded in two stages – the first was
to elicit basic demographic information which would allow a structured
sampling process and the second was a longer in-depth semi-structured
interview. Before the research could proceed it had to be approved by the
Deakin University Ethics Committee. This group insisted on a "Plain
English Statement" of the project which in turn formed the basis of an
informed consent form: a signed statement giving permission for the
interviews to occur, affirming the right of participants to withdraw at any
stage and to have their privacy protected in the storage and publication of
the results. The participants were recruited by gaining the permission of
employment service managers who specifically dealt with women from a
range of backgrounds (differentiated by age, location, ethnic background
and physical ability) and then by using posters and direct invitations from
workers in the agencies. Eventually 38 women agreed to fill in a preliminary
questionnaire about their background and indicated their willingness to be
interviewed for a longer period. A summary of the characteristics of those
women is provided in figure 4.1.

   While in no sense can this group of women be designated and discussed
as a representative sample of unemployed women in Geelong, the 38 who
agreed to participate in the first phase of this project give some insight into
female unemployment in this region. In particular, the profile in figure 4.1
suggests that you are more likely to be an unemployed woman in Geelong
if you are:

- Australian born but with an over-representation of migrants. Thirty-
  four percent of those unemployed were born overseas yet this group
  comprises only 22 percent of the population.
- Over 40.
- Have less than 12 years of education.
- Come from a household with an extremely low income level. 65 percent
  of those who indicated their household income put it at below $200
  per week when the average income for an individual in Australia in
  1999 was c.$650 per week.
- And have been unemployed for more than one year.

What such dimensions meant for the women concerned was explored in
ten in-depth interviews. The schedule asked for the employment history of

each woman and probed the ways in which their ethnicity, age, gender, class, location and physical capacity had influenced that history. As the core question of the project was the perceived and actual effect of such dimensions on employment, the sample was stratified on the basis of age, ethnicity, location, education level, household income and physical ability (see table 4.1).

Central to the project was the question of how the identity of these women was constructed – by themselves, by the employment service agencies, by partners, by the Department of Social Security, potential and actual employers – and how these identities impacted on their job search. This centrality was driven by recent feminist and postmodern literature that posits identity as fluid, contested and contextual (see for example Butler, 1990; Gibson, 1996; Gibson-Graham, 1996). I proposed that such a view could well contrast with what women experience when they report to government bureaucracies, deal with employment agencies, decide who will do the housework and child care and negotiate the job market; in that as a feminist I assumed that gender mattered above all other dimensions of social difference. The main research questions therefore became:

1   What does the experiences of ten women say about the nature of service sector employment in an Australian regional center?
2   How is identity constructed in the job search, by whom and with what effects?
3   How central to identity and workforce location are gender, age, class, ethnicity and physical ability?

The answers to these questions will be explored in detail elsewhere. Here I concentrate on the last question and the light it sheds on the assumptions, or my positionality, as someone doing Feminist Geography research.

Identity Issues: Positioning the Subjects

Looking at how these women see themselves, Cathie and the jointly developed interview schedule asked for a detailed employment history, while a set of specific questions explored how being a woman, of a certain age, ethnicity and physical ability affected the job search and employment. The details of the research, its purpose and an informed consent form were given to all involved. While such a direct approach could be criticized for not probing deeply into the many dimensions which shape work force activity, the responses were revealing and can be read in a transparent as well as a deconstructive way to come to similar conclusions.

What emerged most strongly from the interviews – done in two stages

Figure 4.1 Profiles of unemployed women presenting to agencies in Geelong, Australia, March 1999

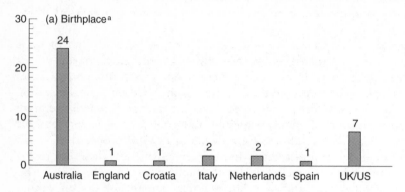

(a) Birthplace[a]

[a] Ethnicity was addressed in a number of ways: by birthplace (registered above), through main language spoken at home (which produced only English), with primary culture identification (which produced a different pattern to "birthplace": Australian 30; English 2; Greek 1; Croatian 1; Italian 1; German 1; and Spanish 1).

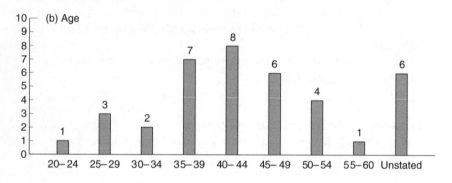

(b) Age

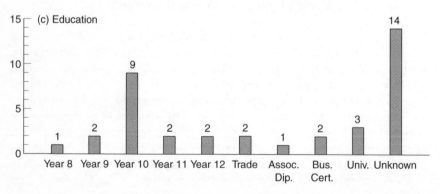

(c) Education

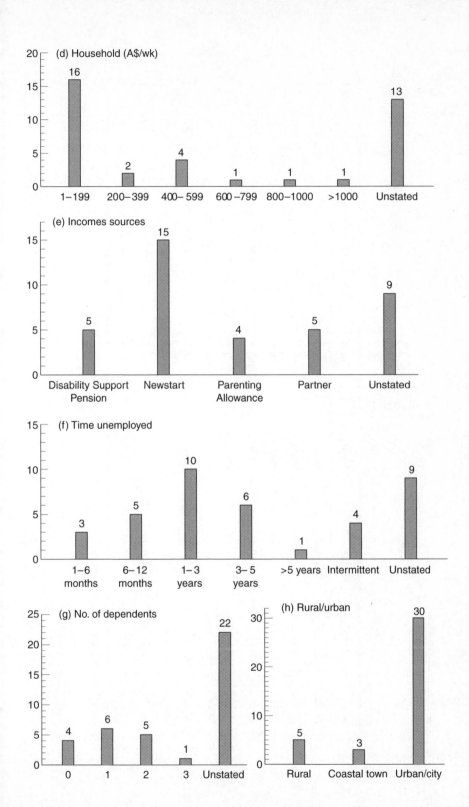

Table 4.1 Respondents in survey of unemployed women in Geelong, Australia, April–July 1999

No.	Age	Birthplace	Rural/Urban	Disability	Education	Time unemployed	Income (A$/wk)
1	45–49	Australia	Rural	DSP[a]	Bus. Cert.	6–12 months	200–400
2	40–44	Australia	Rural	None	University	Intermittent	1–199
3	35–39	England	Urban	None	Trade Cert.	3–5 years	400–600
4	40–44	Italy	Urban	None	Year 11	5 years	200–400
5	20–24	Croatia	Urban	DSP	Year 10	6 months	400–600
6	50–60	Australia	Urban	None	Unstated	Intermittent	1–199
7	35–39	Croatia	Urban	None	Year 10	3 months	1–199
8	40–44	Australia	Rural	None	Year 10	3 years	1–199
9	50–54	Netherlands	Urban	None	Year 10	Intermittent	1–199
10	25–29	Australia	Urban	DSP	Year 11	2–3 years	1–199

[a] DSP refers to the Disability Support Pension

over the space of three months to track the success of their job search – was the importance of being women to their understanding of that experience. So for example, S notes of her second full-time job as a bookkeeper for a solicitor: "The job had originally been offered to a bloke who didn't take it. So it was offered to me at much lower wages. I wasn't worried as I was happy to have a job. Men always got more than women." In answer to the question "How has being a woman affected your job search?" S continues: "Fifty to sixty percent decrease in opportunity. Ninety-nine percent choose a male over a female ... Females don't get a better opportunity. A woman is expected to have word processing and typing skills but a woman also has to pick up children or go home if they are sick. You can pick that up in job interviews. I don't fit that stereotype."

She comments on the complexity of gender roles a few months later as she reflects on her successful bookkeeping business: "I am looking forward to the future. I think the way I live will be directly related to the amount of work I put into my business ... I think small businesses are growing. ... Pay is reasonable. More women are running these types of business. ... We have sold ourselves the wrong message. I would have liked to have stayed home. I don't think it is fair on the child. Women want more in the house, the two cars etc. Children should give you a sense of self. Men just expect women to have multiple roles."

S also expected multiple roles of herself – of businesswoman, mother, homemaker and woman of independent means. However, all of these dimensions are based on her sense of self as a woman who has been treated in a particular way within the job market and at home because of her sex.

So too with V, who has done a range of jobs – including working at a service station, for a security firm, as an accounts clerk, census collecting and doing reception work and house cleaning – but who sees herself primarily as a mother. It is with this role that she identified and around which her paid work had to fit. She tells her story: "I left school when I was sixteen because I had a job ... in an office doing accounts. ... I had that job for 11 years ... then I got a part-time job with a security firm – cash accounting, making up payrolls. I left this job after 18 months as I was pregnant ... I worked in an office of the service station during the school holidays and then did a cleaning job ... I wouldn't travel to Melbourne and I don't want a full-time job because of [my] kid's needs."

Often when asked the direct question on whether being a woman mattered in their job search and experience, women replied "No," but then frequently went on to contradict their own judgments. For example R said: "I don't think being a woman has affected my job search. I don't know whether it would be any different from a woman or a man. Because I've been in machining and now in admin. ... It's a woman's area. There has been no challenge to be against blokes."

And G, who had been in the military, teaching, on a newspaper and in security observed: "I can't see that being a woman has any effect on my work. The only time [was when I went] to a security firm. I have a shooter's license so I met the requirements. All they wanted was a communications worker . . . but I was knocked back for being a woman. They gave all sorts of excuses, but I met all of the criteria."

So too with B who is a trained teacher working part time with the elderly: "I don't really know how being a woman has affected my job search. I don't really know that it has had a lot of bearing – probably because of the type of work I'm looking for is within a woman's field."

Complicating but not overwhelming gender – in fact adding to its potency – are the dual issues of parenting and age. And the importance of age is most evident when a woman turns 40. S notes: "Hospitality work is easy to get but for younger people. People under 40 get jobs and males against females. . . . [You are] told that it will be a plus [to be over 40] but I haven't found this. Employers often take looks into account. A younger person looks attractive. A lot of time looks come into it. They like the trendy young ones. A lot of the employers, even the employers are younger than me. I have gone to interviews where I am ten years older than all of them. If I was the boss I would probably pick the young one. . . . Age has affected my career. We are living in a time where retirement age is lower. In restructuring people of a certain age get shoved aside. We should be considering older people; they have the experience behind them."

This is also the case for R who has worked as a machinist and has been looking for clerical work for three years: "I think once you're over 40 and a woman they think, 'what's the point?' . . . Once you're in your 40s you are over the hill. I really feel like in the elderly group. I think it should be better when you're in your 40s – you don't have young children, or the hassles of young life and you're more in tune. Sometimes they like younger women because they're more attractive and thinner."

There were also the issues of ethnicity, the migration experience, education type and level, but these, too, for the feminist researcher can be related to gender. So for H: "I came from Holland and didn't go to school until I was seven years old, when I came to Australia. Until I was seven I kept the house. My brother went to school but girls stayed home. . . . If you have an education it is easier to get a job. In my day you only got a little choice – cooking, sewing, secretarial, Coles or Woollies or waitressing. The education I received at school was not good for job search. That type of education is only good for getting married. . . . I don't think being a woman has affected my job search. There are plenty of men out of work."

For T, being born overseas was critical to her chances of getting work, as an accent in her English continually went against her. She also left school at 15 "because my mother told me that education was wasted on girls" and proceeded to work in secretarial and clerical positions before migrating. On her job search she notes: "It's easy to get a job in fast foods because of the cheap labor. Receptionist jobs are hard to get. I think my accent is a problem. I was told that nobody understands me. People with the gift of the gab get the jobs. They can talk their way into them; it doesn't mean they can do the job. . . . I once had a mock interview and he told me to change my accent. He said I didn't pronounce my words correctly."

While a small sample, my analysis of these words confirms the importance of gender in the self-perception, identity and work force activity (and frustrating inactivity) of these unemployed women. Combined with their location in a particular part of the service sector – in the lower end, part-time casualized part of it – their experiences affirm the extraordinary level of economic and social inequality, which still pervades Geelong, and, I would argue, the rest of Australia.

The Difference Feminism Makes

As a socialist feminist researcher my focus is on women, while my key assumption is that the patriarchal economy – that system of labor expenditure, appropriation and reward which tends to systematically advantage men over women – shapes women's experiences of unemployment and re-employment. In this project, research design – the methods used, the choice of research assistant and respondents, the questions asked and the analysis – followed from this assumption. However, I am also a feminist highly engaged with the postmodern and postcolonial critique of socialism and identity politics. As such I accept the importance of considering multiple self-definitions and the complex constitution of identity. Together, these assumptions led to the design of this project – to explore the relative importance of gender, ethnicity, age and disability on women's employment histories. However, positioning myself primarily a feminist, the research design and the interviews have led to one conclusion: that a focus on women exposed not so much their multiple and fractured identities but the importance of gender and women's ongoing oppression. The conclusion that gender is the main determinant of women's labor market experience is indeed the difference feminism makes, but it also raises some profound questions about the nature of my feminist research standpoint. I would argue that my feminist position, choice of research assistant and research methods predetermined this outcome and need to be radically reassessed to admit the possibility of being proved wrong! Postmodern feminism has to acknowledge that gender may not always be the main determinant, while techniques – quantitative and qualitative – must allow the real possibility of gender being relatively unimportant. For in the case of the interviews described above, a serious admission of ethnicity, disability and age into the explanatory framework could have produced a very different conclusion.

So, for example, consider the experience of B, the most highly educated of the unemployed women talked to. She completed high school and trained successfully as a primary school teacher – a career choice made by many more women than men; a highly gendered decision. She then had a number of years teaching and began postgraduate study. However, her studies had to be deferred owing to "family circumstances." Shortly thereafter, a series of untreated ear infections led to partial deafness and initially to her leaving work in schools and, as the problem worsened, an end to current work with the elderly and attempts to break into hospitality. Not only is the disability proving a hindrance to her identification and pursuit of employment opportunities, but so too is age and a further accident which has impaired lifting ability. She tells her story:

> I was at HC from 1970 to 1995 – I had junior primary and then I did work as a remedial support teacher from prep to grade six. I left because my hearing had dropped. I then did remedial work but the funding was cut . . . It's not that I can't each anymore. It's that running a discussion group had become difficult because of my hearing loss. . . . I don't really know how being a woman has affected my job search. I don't really know that it has had a lot of bearing. Probably because of the type of work I'm looking for is within a women's field. My age has had a lot more bearing, because I think that some employers are wondering why I am working at my age or if I was going to leave. . . . Disability? That has not actually been said to me at an interview, but with some part-time work it has come up. They never ask me in an interview – it is in my covering letter about the loss. Having the hearing loss has made the interview situations more difficult.
>
> [Three months later after an accident which signaled osteoporosis] I can only do light duties. Looking for a job will now be much harder because of my health . . . Many of my options have closed, but I would like to complete the course and tie up what I am doing. My capacity to look for work has changed.

So for B, while gender has clearly shaped her educational and employment choices, it is now primarily issues concerned with her age and illness that are shaping her employment history. It is a postmodern story of multiple identities and oppressions with gender not at its center all of the time. But yet it is a particularly female story; one perhaps best apprehended by a researcher with multiple identities as a socialist, postmodernist, and feminist!

STUDY MATERIAL FOR TAKING ON FEMINIST RESEARCH

WORDS

Binary thinking
Colonialism/postcolonialism
Context
Empiricism
Epistemology
Feminism
Feminist geography
Gender relations
Humanism
Legitimation
Masculinism
Methodology
Narrative
Objectivity
Oppression
Patriarchy
Positivism
Power
Privilege
Representation
Second wave/third wave feminism
Situated knowledges
Standpoint theory
Text
Validity

QUESTIONS

In what ways can you see feminist research practices transforming geography? What possibilities might feminist geography bring to feminism?

Develop a working definition of feminism. Of feminist geography. Of feminist research in geography. Revisit your definitions and pull out truth claims about knowledge. Is it possible to identify any particular type of feminism? Feminist epistemology? Feminist praxis?

Several of the authors make clear that they see an intrinsic contradiction between feminism and the system within which academics work. In what practices do you see this contradiction emerging? What sorts of resolutions are you comfortable with?

Think about ways that your multiple positionings have affected your learning experiences in the classroom, in doing research, and in applying your knowledge. What ways have you witnessed forms of knowledge and social relations being legitimized? Invalidated? Accepted? Challenged? Reproduced? Contested? Destabilized?

ENGAGED EXERCISES

Manifestations of power and knowledge

Think through how productions of power and knowledge manifest themselves in specific places and record your general impressions. Then draw a map of places you frequent or mark a map of your campus, your home, your workplace, or your home town and identify the various intersections of power influencing these places. Consider factors such as how you access, navigate, and experience these spaces, the contextualized meanings each space evokes, and what sets of power relations affect your reading.

Share your maps and interpretations. Are there common themes emerging about how power and knowledge manifest in the various places mapped? In what ways is each map unique? What can you say about the manifestations of power and knowledge in specific spaces and places?

Comparing types of research

Create a table with a grid of five cells across and eight cells down. Across the top, beginning in column 2, list various orientations of research, as for example, feminist, marxist, positivist, and poststructuralist. Or, perhaps list various types of feminist research, as for example, socialist feminism, postructurally informed feminism, second wave feminism, and empiricist feminism. Down the side, beginning in row 2, list aspects of the research process that are influenced in some way by the choice of research orientation. For example, list truth claims, types of research methods favored, sources of inspiration, types of research questions that are asked, research employment and mentoring practices, types of political action, and any other aspect that you find significant in the research process.

Fill in each cell according to the orientation of the research (across the top) and the aspect of the research process affected (down the side). With a learning partner, discuss the inclusions, exclusions, absences, overlaps, contradictions, and paradoxes in the cells.

SUGGESTED FURTHER READING

For an nice introductory overview of methods and methodologies in feminist geographies read the chapter in *Feminist Geographies* (Madge et al., 1997). Good places to start in trying to understand the ways in which feminist geographers are engaging methodological issues within feminist geography include: the short edited theme papers appearing in *Canadian Geographer* (1993), *Professional Geographer* (1994, 1995, 2001), and *Antipode* (1995); the "Methodology" section of *Thresholds in Feminist Geography* (Jones, Nast and Roberts, 1997b); and the methodology chapter in Linda McDowell's book, *Gender, Identity and Place* (1999).

Arising out of the study of the politics of the production of knowledge are issues about the challenge to Cartesian dualisms, power and privilege, and feminism in physical geography. Although somewhat dated as a topical review, Liz Bondi (1992) addresses gender and binary thinking in feminist geography. This piece is useful in understanding *how* feminists are thinking about how binary thinking affects scholarship. Gillian Rose (1993) proposes paradoxical space in *Feminism and Geography* as a way to understand and work through dichotomies. Vera Chouinard and Ali Grant (1995) talk about power and privilege within feminist geography and how disabled women and lesbians have been "left out" of the feminist project. Gillian Rose (1997) critiques feminist geographers' attempts to date in the literature on dealing with power within the research process. Clare Madge and Anna Bee (1999) argue that gender is a considerably important set of relations shaping how physical geography exists within the academy. Sheryl Luzzadder-Beach and Allison MacFarlane (2000) look at the differences in the practice of physical geography for women and men. The articles in *Professional Geographer* (2001) look at how women in geography have struggled to be there.

There are a number of feminists writing about feminist epistemology in the social sciences and humanities that are shaping ideas about the politics of knowledge production within feminist geography. One key piece is Donna Haraway's (1988) article on situated knowledges. Patricia Ewick (1994) shows how to integrate feminist epistemologies into undergraduate research. Parvati Raghuram, Clare Madge and Tracey Skelton (1998) discuss the pedagogical implications of doing feminist research in under-graduate geography courses. Some other writings we found useful in looking at feminist methodology generally in our studies are: Harding, 1987a, 1991, 1992; Fonow and Cook, 1991; Alcoff and Potter, 1993; Wolf, 1996; McDowell and Sharp, 1997; and Devault, 1999.

Finding papers of interest in geography journals is another way to enhance familiarity with the ways feminists approach various topics. Perusing the pages of *Gender, Place and Culture, Antipode, Environment and Planning D: Society and Space, Environment and Planning A,* and the *Journal of Geography in Higher Education* might give you more concrete ideas on how to set up a feminist geography research project.

Part II

Thinking about Feminist Research

Delimiting Language?

Feminist Pedagogy Working Group

When thinking about feminist research, there seems to be no escaping from pages and pages of text, saturated with the dense language of theory. Arguments ensue as to whether such practice is liberatory or elitist. On the one hand, is there a problem with feminism as a social movement when, because of feminist theory's seemingly indecipherable vocabulary, so many of its texts exclude all but a privileged academic elite? Or, on the other hand, are "new" terms like "subjectivity" and "positionality" – terms that become meaningful only after one understands their context within a larger academic debate – invaluable tools that allow us, as feminists, to imagine new ways of understanding ourselves and society? After long, detailed discussions, drawing on experience in feminist organizing and several (types of) academic degrees, our collective has concluded that the language of contemporary theory is both emancipatory and confining, offering possibilities while at the same time limiting our choices. In writing this piece, we want to occupy this space of ambivalence: we want both to defend the language of theory, which we feel has been integral in allowing us to grow as thinkers, while acknowledging that its abstruseness and consequent exclusivity are indeed problematic.

At times, theoretical language sets up walls of exclusivity. In the drive to create ever more specialized terminology to articulate previously neglected experience – the experiences of women, of people of color, of people with different sexual orientations, and of other groups of people made "different" through being marginalized by some social division of power – it is ironically those whose experience initially provided the impetus to expand language who are often relegated to the realm of the "theorized about." At its best, however, theoretical language provides tools for change. Feminist research reminds us that while at times it is necessary to push theoretical boundaries, we must also remain committed to the material, always conscious of our study as research *for*, rather than merely *on* human beings.

The language used in this book illustrates this tension between political emancipation and theoretical elitism. In discussing each of the articles in the following section, for example, different members of our collective expressed varying levels of comfort with the way in which each of the authors used theoretical language. Some argued that geography as a discipline has been enriched by contributions of feminist scholarship with the addition and use of terms ranging from "paradoxical space," to the developing concept of "reflexivity," Yet questions lingered about whether or not the authors could have conveyed their ideas just as clearly without using abstract or alienating terminology. A term such as "subjectivity," for example, continually surfaces in different guises throughout the chapters in this section, making it an extremely difficult concept to come to terms with. This could leave one wondering if the variances in meaning end up clarifying or merely obfuscating what the authors are trying to express; and yet, few other combinations of simpler words can convey such an array of meaning with the elasticity and succinctness of this singular term.

To those of you who approach this new language with a healthy suspicion, we can recommend only that you immerse yourself in it, if only for a short while, before you decide whether to embrace or reject it. The project of gaining fluency in "theory-speak" may be made easier if you change your expectations surrounding how long it should take you to read an article: if you normally read a page of text in ten minutes, expect to take an hour. As you undoubtedly have noticed, your standard college dictionary may not be of much use in helping you puzzle out the meaning of specialized theoretical terms. We recommend that you purchase or gain access to a dictionary of specialized geographical terms such as *The Dictionary of Human Geography* (Johnson et al., 2000) or *A Feminist Glossary of Human Geography* (McDowell and Sharpe, 1999), which will allow you to contextualize terms and ideas within a broader debate. Remember that entering into theory is akin to walking into a conversation that has been going on for decades and has developed its own points of reference, inner debates, and, sometimes, "in-jokes," so expect that your learning curve will be quite steep as you make your way through your first few articles.

On a less practical level, as students with our own feelings of ambivalence about theory and its language, we recommend that you approach language in theory not as unalterable dogma, but rather as a process to be negotiated. The terminology of negotiation affords a more subtle, flexible approach to the relationship between feminism and language. To negotiate something implies not only a certain artful finding-of-one's way through a problem, challenge, or dilemma, but also an engagement in dialogue with others to reach a compromise, consensus, or agreement. A brief exploration of the word's history lays to rest any hesitancy towards compromise as either a sell-out or a quick fix, for the Latin roots of *neg-* (meaning "not") and *otium* (meaning "leisure") remind us that the journey will not be an easy one. This does not mean that the journey is not worthwhile; in fact, the difficulty associated with voicing some misgivings about language may be the very thing that makes our efforts valuable, for it is out of this sense of discomfort that learning, struggle, and action can arise.

Topically, the chapters in this section focus on identity, subjectivity, positioning, and the practice of feminist research. They introduce new terms, conceptualize processes in new ways, and provide new insights into the ongoing discussion of how to go about thinking about feminist research in geography. David Butz and Lawrence Berg, in their proposal to use duppy feminism as a metaphor for men's places in and masculine position-ings within feminist geography, take up the notion of paradoxical space as a place from which to engage in practical action as well as an uneasy grounding in a particular subject positioning. Karen Falconer Al-Hindi and Hope Kawabata attempt to work through how a feminist researcher can become more fully reflexive, not only at the time of interaction with research participants, but also as feminist subjects within geography. Gill Valentine suggests that putting theory into practice involves more than being sensitive to the research process; it entails moving toward interpreting various performances of identity, both the researchers' and the research participants'. Together these chapters cover a wide range of issues about processes constituting difference, subjectivity, and identity that have been and will continue to be debated in feminist methodology in geography.

Short 2 Putting Feminist Geography into Practice

Gender, Place and Culture: Paradoxical Spaces?

Liz Bondi

Feminists and feminisms have always occupied uneasy spaces within the academy. For some, the very idea of a feminist academic is a contradiction in terms because such academic necessities as taking up a position of academic authority – one whose knowledge is valued and privileged in a distinctive way – militate against the anti-hierarchical, anti-elitist egalitarianism integral to many versions of feminism (for critical discussion see for example Bondi, 1997; Friedman, 1985; Hawkesworth, 1989; Morgan, 1992). More generally, feminists have amply demonstrated that academic knowledge and academic practices are riddled with ideas and assumptions that depend upon and generate gender inequalities and biases (for classic statements about science, philosophy, and geography respectively see Harding, 1986; Lloyd, 1984; Women and Geography Study Group, 1984). Consequently, for women and men to acquire academic knowledge and to perform successfully within the conventions of academic practice, we are required to participate actively in the enactment and reproduction of gender inequalities and biases. A couple of examples will illustrate the impossibility of doing otherwise. First, if I wish to advance a feminist perspective on any academic debate you choose to name, conventions demand that I demonstrate my understanding of existing contributions, and that I give serious attention to texts widely recognized as "important" and "weighty." In other words my contribution must be situated in relation to existing traditions, however steeped in misogyny, androcentrism, or gender-blindness these may be, and I must, in effect, restate the importance and weightiness of these traditions if my attempt to question them is to be taken seriously. Second, if I attempted to challenge every instance of gender bias I observe in my working life, I would never rest, let alone get on with the work I am paid to do. In other words, I have to compromise my commitment to feminism in all aspects of my work including the most and the least routinized.

The word "compromise" has acquired both positive and negative inflections. Positively, "compromise" is understood to entail a willingness to settle differences "by mutual concession" (*Chambers Dictionary*); negatively, it is understood as a neglect leading to "risk of injury, suspicion, censure or scandal" (*Chambers Dictionary*). This doubled quality may help to explain why the position of feminist academics is so often described as contradictory, that is as "inconsistent" or as encompassing "two propositions that cannot both be true" (*Chambers Dictionary*).

This account of the position of the feminist academic prompts the question of why any of us do it! Such a question could be approached in many ways, but what I want to suggest is that the impact of these various compromises and contradictions depends upon whether they operate as binary (mutually exclusive) oppositions or as paradoxes, that is as "self-contradictory statement[s]" which may appear to be "absurd" but which might nevertheless also be "true" (*Chambers Dictionary*). Clearly if on the one hand "feminism" and "the academy" always operate as mutually exclusive oppositions those of us who attempt to straddle the two will forever be pulled apart. If, on the other hand, "feminism" and "the academy" operate within the framework of paradox, then their uneasy relationship might contain possibilities for "absurd" surprises and associated pleasures.

My argument draws directly on the practices of one feminist academic journal, namely *Gender, Place and Culture: A Journal of Feminist Geography*, which Mona Domosh and I founded and for several years edited. Preparatory work for the journal began in 1992 and we swiftly secured the interest of publishers Carfax, now part of the Taylor and Francis Group. The first issue was published in 1994. Mona stepped down as editor after the first four volumes and I continued for another two years, by which time a new pair of editors – Lynn Staeheli and Gill Valentine – had taken over. In this chapter I discuss some of the compromises that influenced the development and character of the journal, suggesting that these might be viewed as illustrations of feminist efforts to operate both "within and against" the academy. This inconsistent and contradictory positioning can make life difficult, and on many occasions the conflicting demands of academic conventions and feminist values deeply felt by a few individuals leaves the academy completely untouched. But I suggest that if the conflict between what it takes to operate within the academy and what it takes to argue against it can be held as a paradox, then efforts to work both within and against the academy can be fruitful. Thus, I argue that at its best *Gender, Place and Culture* can be understood as an embodiment of "paradoxical spaces" in which feminist geography is practiced creatively and productively. But I will also suggest that the journal's capacity to operate in this way is always fleeting, uncertain, and contestable. Through this particular focus I hope to point to some more general features of feminist academic practice including its methodology.

Gender, Place and Culture established an alternative to existing geography journals, offering instead a space dedicated specifically to feminist perspectives. When the project was first mooted, my conversations with colleagues rehearsed the pros and cons of such a strategy. Among the dangers we considered was the risk that the journal would work against the transformative challenge of feminism by reducing the flow to other geography journals of manuscripts informed by feminist perspectives, and by limiting feminist work to a ghetto that others would ignore. At the same time, there was the possibility that the journal would consolidate the place of feminist perspectives within the discipline, render feminist work more visible within and beyond the discipline, and thereby enhance the scope for feminism to challenge and transform geography.

In proceeding with the project those of us involved felt that the potential benefits outweighed the risks. I do not wish to evaluate this judgment in relation to subsequent events but instead I want to suggest that these informal discussions drew on a particular way of looking at questions about feminist academic practice, a framing characterized by a distinction between establishing alternatives to an already existing "mainstream" and transforming current practices from within the "mainstream." This framing was certainly familiar. For example, an early debate about the development of geography curricula inclusive of feminist perspectives distinguished between the introduction of new courses concerned with the geographies of women and of gender, and the revision of existing courses to incorporate critical awareness of women's lives and gender inequalities, classifying the former as separate provision and the latter as integration (see for example McDowell and Bowlby, 1983; Monk, 1985; Peake, 1983). We recognized that each of these strategies has advantages and disadvantages. Separate provision had the advantage of creating new kinds of opportunities for interaction between teachers and students with feminist leanings, often in classrooms consisting wholly or largely of women. But this approach had the disadvantage of limiting the impact of feminist perspectives to a single course and a self-selected group while the rest of the curriculum and the rest of the student body was left unchanged and unchallenged. Conversely, integration had the advantage of carrying the influence of feminist ideas to a much wider audience and promised a radical transformation of the whole of the curriculum. Against this, integration suffered the disadvantage of leaving feminist teachers and students isolated from one another. The approach also ran the risk of the radical potential of feminist ideas being neutralized by routine references to "gender" as a (largely unexamined) social category without any attempt to address the conceptual challenges associated with feminist perspectives (see Bondi, 1990a; Christopherson,

1989; and Johnson, 1989, for fuller elaboration of these processes). These pros and cons applied in similar ways to questions about a journal dedicated to feminist geography: would the existence of an explicitly feminist geography journal encourage the development of innovative and influential feminist work within geography, or would it create a ghetto in which feminist scholars spoke only to one another?

Such questions were, of course, unanswerable, and the decision to set up the journal illustrates the point that feminists make choices about how to *do* feminist geography without knowing what the effects will be. Since we care a good deal about those effects it is hardly surprising that feminist academic practice often feels uneasy. However, I want to suggest that the unanswerability of such questions is also, paradoxically, conducive to the creative and fruitful practice of feminist geography.

The strategies of making separate provision for feminist work and of integrating feminist perspectives into the mainstream are not, in practice, mutually exclusive oppositions. Thus, it is possible to offer specialist courses and to include feminist perspectives within "core" courses within a single program (compare Monk, 1985). Likewise most of those who publish in *Gender, Place and Culture* also publish in journals that are not explicitly or exclusively devoted to feminist perspectives. We might argue therefore that the decision to set up *Gender, Place and Culture* did not and does not restrict feminist geographers to one strategy at the expense of another. This does not mean, of course, that the existence of the journal makes no difference to the character of feminist geography or to its impact on the discipline. Rather, since these strategies are not mutually exclusive, feminist geographers can work with "both/and" possibilities rather than "either/or" choices. From a "both/and" perspective, we can use the distinction between separate provision and integration to clarify the differences between different strategies without being forced into a direct conflict between contradictory principles. And it suggests that many feminists are interested in working both "within" the academy – by integrating feminist ideas into "mainstream" outlets – and "against" the academy – by creating spaces which challenge the limits of, the "mainstream."

So our choices make a difference but are rarely if ever unambiguous in their effects. Working with an awareness of their ambiguous consequences, for example by revisiting the pros and cons of different strategies, works instead to encourage creative ways of both adhering to and contravening academic conventions. This begins to illustrate what I mean by working within a paradoxical framing of the relationship between feminism and the academy. I would also suggest that the asking of unanswerable questions is productive. In this instance it heightened awareness of the impact of our decisions and in so doing it may have helped to ensure that, collectively, feminist geographers sustained both the strategy of separate provision and the strategy of integration. One of the broader issues to which this points is that posing unanswerable questions may often be a useful methodological strategy. This runs counter to influential conceptions of science which claim that decisive testability is the hallmark of good research questions (for the classic statement see Popper, 1959). But if our aim is to influence the world as well as to study it, questions that sustain critical thinking in relation to academic practices may turn out to be more effective.

Some decisions we made when we set up *Gender, Place and Culture* were of an altogether "harder" variety (and the phallic associations of the adjective are not irrelevant) in that they required us to choose between mutually exclusive possibilities. For example, we had to set up a structure within which to make editorial decisions. A key question we faced was whether we should entrench feminist principles of collaborative working by setting up an editorial collective or whether the journal should adopt the "mainstream" practice of having one or two named editors together with an editorial board. In choosing the latter we were well aware of the pros and cons of both, for example the potential that collective working would make for slower progress as against the risk of creating something in which only a minority of feminist geographers consider themselves to be involved. (Contrast this decision with the collective approach adopted by such journals as *Feminist Review, Feminism and Psychology*, and *Frontiers*.) Several other decisions also led to the adoption of conventional practices rather than alternatives, for example in setting up arrangements for reviewing manuscripts, where we adopted a traditional double-blind system (see Bondi, 1998).

It might reasonably be argued that the structures adopted by *Gender, Place and Culture* conform very closely to "mainstream" academic practice: beyond the subtitle of the journal there is little if anything to suggest any significant departure from, or challenge to, the pre-existing norms of refereed journals. Some might argue that this does not matter; all that matters is that the content of the journal fulfils its explicit commitment to be a journal of feminist geography. But feminists have long insisted that the means influence ends so that how people (authors, referees) and material (written texts) are treated in the production of the journal necessarily affects the end result. Given the relatively small size and highly interconnected nature of academic communities it is likely that the views of prospective authors about how they are treated will be communicated to potential authors and so influence the eventual content of the journal, suggesting that the practices of the journal are of considerable significance. So how should the adoption of conventional structures be understood?

Did we compromise (in a neglectful sense) feminist principles in pursuit of academic acceptability? There are some indications that we did. For example, the journal has been criticized, informally if not in print, for adhering to and reproducing conventional definitions of academic standards in which "clever" and often abstruse theorizing is valued more highly than other forms of writing. This, critics argue, sustains familiar exclusionary practices including a very narrow meaning of "international," which largely excludes feminists from outside Western Anglo contexts and especially those based in the so-called "Third World." In addition, those involved in the editing process are sometimes made painfully aware that those we interact with – authors and referees – can be quite hostile to attempts to deviate in any way from standard practices. And when the founding editors began the process of finding replacements we became more sharply aware of the way in which the reputation of the journal had been attached to us as individual academics rather than being perceived as belonging to a community.

There are also indications that we might have created something more paradoxical, or, to use the definition offered above, something "absurd," in the form of a journal that embodies both highly conventional academic practices, and, contradictorily, inescapably feminist practices. For example, while the manuscripts we received included many that took a form no different from those submitted to "mainstream" journals, there has also been a steady flow of material of a more experimental form, some of which has made it through to print (see for example Reichert, 1994; Hurren, 1998; Okoko, 1999, together with Robson, 1999; Jarosz, 1999; and Laurie 1999). Likewise, notwithstanding the rejection of some manuscripts, those submitting papers have generally received a good deal of encouragement sometimes via the generous reports provided by referees and sometimes from the editors themselves. This has led to the publication of a substantial number of excellent articles by postgraduate students and other "less experienced" authors. More generally I would argue that it is in the communications between editors, referees and authors that the journal has made a small, uneven, but potentially significant contribution to modifying academic practices. To claim that such shifts in practice create sufficient self-contradiction to be viewed as paradoxical may verge on the grandiose, but if feminism is vital to our practices as well as to our concepts then it is in these unpublicized and apparently routine activities that its impact is likely to be felt. The connection with wider methodological concerns is clear: feminist research practice requires that we attend carefully to the relationship between those positioned as "researchers" and those positioned as "research subjects," "informants," "interviewees," and so on. Rarely, if ever, are the decisions we make clear cut, but unless we think about these details we will undoubtedly reproduce dominant forms of knowledge production.

Drawing on Gillian Rose's (1993) discussion of the politics of paradoxical space, Caroline Desbiens (1999) has argued that efforts to reach "beyond" the limits of geography may risk removing feminist challenges from the discipline as it is known and practiced. Both Rose and Desbiens offer tools for thinking and therefore producing knowledge in ways that subvert from within. My contribution here complements such efforts by focusing on some of the "nitty-gritty" activities that make up academic practices. Processes of journal publication are an integral part of these practices.

RESEARCH TIP

Feminist Relationships

- Accept that relationships will not always be harmonious.
- Maintain awareness of complicity in relations of oppression.
- Dispense with expectations when assumptions surrounding rapport dissolve.
- Remember that difference and sameness can be both challenging and rewarding.
- Be sure to have fun!

5

Paradoxical Space: Geography, Men, and Duppy Feminism

David Butz and Lawrence D. Berg

In this chapter we attempt to trace some of the contours of the often fraught – and certainly controversial – relationship(s) between men and feminist geography. In particular, we wish to examine some of the ways that men who participate in feminism as feminist geographers are positioned by contradictory and problematic discourses of masculinity, femininity and academic knowledge production. We draw on poststructuralist perspectives in order to deconstruct binary distinctions between male/female (as ostensibly "natural" embodied positions within the sex/gender system) on the one hand, and masculinity and femininity (as socially constructed subject positions for men and women) on the other. Like Judith Butler (1990), we claim that the distinction between sex and gender is a social construction that produces the very effect it claims to describe. The sharp distinction between sex and gender is a logical result of what Butler calls a "regulatory fiction" that is naturalized through "performativity" – a consistent repetition of behaviors that create an illusion of a natural and "real" gender distinction. Therefore, rather then separating sex from gender, we treat them as part of the same system of patriarchal and heterosexist domination.

We have developed the concept of "duppy feminism" as a key part of our discussion of men and feminist geography. The phrase "duppy feminism" is adapted from the term "duppy," which is used throughout the Caribbean to describe a variety of sly and malevolent ghosts (Johnson-Hill, 1995). The notion of the duppy helps us to think about the malicious ghosts of masculinism in ways that resonate with our own experiences of both contesting and unwittingly reproducing masculinism and sexism within geography. We feel that the ideas of duppy feminism aptly describe the position of male academic geographers who are at once committed to the theoretical, philosophical, and practical tenets of various forms of feminism informed by poststructuralist thought, yet who at the same time often fall short in fulfilling these tenets in our own work and actions. In developing the notion of the duppy feminist, we could be accused of constructing yet another theoretical approach for men to participate in feminist geography – thereby (re)constituting a longstanding theory/empirics binary that articulates masculinism in geography (see G. Rose, 1993; Berg, 1994). This is neither our intent nor, we hope, the way our discussion of duppy feminism will be received by readers. Duppy feminism is, for us, not so much a "theory" as an especially evocative metaphor to help us think about our own problematic and contradictory participation in feminist geography.

This chapter, then, is our attempt to illustrate some of the contradictory positioning(s) of men who work within feminist geography. We argue that such paradoxical spaces (from G. Rose, 1993) are a necessary and productive starting point for men to participate in – and more importantly, to support – feminist geography.

Men, Masculinity, and Feminism?

We begin our discussion of men and feminist geography with two stories that we think help to highlight the contradictory positions that people find themselves in when attempting to effect a feminist subjectivity within geography.

The first story begins at a recent conference where we both participated in two special sessions in which the topic of discussion was feminist qualitative research. These sessions brought to the surface, once again, many of our insecurities about participating in feminism. For example, we were both uncomfortably aware of our own male bodies intruding onto the scene – immediately calling into question our credentials as feminist commentators. Our insecurities as "feminists" led us both to take much of our respective presentation times to position ourselves within the various discourses of feminism. Interestingly, of the eight other participants in these sessions, only one other person took the time to explicitly position herself in relation to feminist geography – and she did so in order to make it clear that she "was not a feminist." While these quite minor events might be read in many ways, we both took them to mean that there was an implicit assumption among at least some of the other participants that their female bodies signified an automatic link between themselves, their comments, and feminist geography.

The various epistemic spaces in which the people participating in these sessions found themselves suggest a geographically specific and culturally contingent juncture in the production of feminist geographies. In this sense, they point to a particular epistemological understanding of feminist geography that draws on liberalism for its inspiration. Here, liberalism allows us to theorize a direct and unproblematic link between female bodies and feminist subjectivities. Interestingly, few of the participants would consider themselves to be solely liberal feminists and this in itself illustrates the effect of liberalism. In this situation and many others like it – even if we consider ourselves to be socialists or poststructuralists – most of us cannot escape the materiality of bodies nor their (liberal) theorization.

Liberalism, whilst an important component of contemporary Western feminisms, offers certain kinds of opportunities while posing other kinds of constraints on the participation of men in the feminist "project." There are other approaches – especially poststructuralist-inflected feminisms – that provide different and perhaps more productive opportunities for men and others to work within the interstices, contradictions and paradoxical spaces of everyday life.

This leads us to our second story, one that focuses on the difficulty that we had in developing this essay and in so doing, traces out our own contradictory positioning as pro-feminist men in geography. Our story begins with our initial reticence at accepting the invitation to write a paper about men and feminist geography. Neither of us felt all that comfortable about the prospect of writing this essay; nor were we comfortable about how it might be received among our peers in the wider community of feminist geographers. At the same time, we felt that the topic of the paper – men and feminist geography – was too important to not include in a collection of essays about feminist methodologies in geography. Our concerns about the project were not lessened when we eventually got up the courage to (partially) confront our own contradictory participation in feminist geography. Our first attempt at writing an essay about men confronting masculinism in geography was, in some senses at least, a complete failure. Indeed, instead of contesting masculinism, it could be argued that we managed to (re)produce a very masculinist text. The subsequent revisions have not been wholly successful in expunging the "ghosts" of masculinism from our text. On reflection, however, that may not be such a bad thing for our readers, as some of the contradictions of our text may help to illustrate in a tangible way the very contradictory and paradoxical character of our own participation in feminist geography.

Both sets of events we describe above also point implicitly to what Sandra Harding (1998, p. 172) refers to as a "hesitancy" among female (and other) feminists to accept the work of heterosexual male feminists (or pro-feminists) at face value. Certainly, there are good reasons for such hesitation: "The deepest, most widespread, and most influential forms of sexism and androcentrism are neither overt nor intentional, but, rather, institutional, social, and 'civilizational'" (Harding, 1998, p. 173). In this sense, sexism, masculinism, and patriarchy are so pervasive in the West, that it is difficult even for feminist women geographers to identify and theorize some of their more invidious aspects. In addition, men, who are the (usually unknowing) benefactors of patriarchy and sexism, have a much more difficult time identifying the sources of their own male privilege in society.

Yet there are many reasons to reject the uncritical acceptance of a stance that *automatically* questions the ability of men to participate in, and contribute to, feminism. There have been many men who have helped us understand the sexist character of Western societies. The list of such men includes academics like John Stuart Mill and Karl Marx or more recently Harry Brod, Robert Connell and Michael Kaufman, or more popular figures like Woody Guthrie and "political" leaders like W. E. B. DuBois. A number of male academic geographers (see for example Jackson, 1991, 1993; Sparke, 1994, 1996) have also contributed significantly to feminist understandings of geography. Perhaps more importantly, however, a feminism that refuses to accept the possibility that men might be active participants in the transformation of masculinity is both practically and theoretically problematic.

In practical terms, if we subscribe to a discourse that refuses the possibility that men can change themselves then such a feminism is doomed from the start, for we will be constructing a monolithic and immutable masculinity that cannot be transformed by feminist praxis. In theoretical terms, such a feminism fails to understand the way that socially constructed hegemonic masculinities work to disempower women and subordinate other men. As social constructions, masculinities are heavily influenced by hegemonic cultural forms. These forms constitute the set of values that define what it means to "be a man." They tell men what is acceptable in terms of dress, conduct, values, aspirations, body shape, tastes, and desires, if they are to be seen as suitably "masculine" in social life. However, the demands placed on individual men by these hegemonic cultural norms are not uniform. In fact, in their articulation with different (and sometimes contradictory) aspects of everyday life, they are more often than not fractured and contradictory. The achievement of a unified masculine identity is thus a virtual impossibility. Nonetheless, there does exist a hegemonic masculinity that defines a narrow set of gendered possibilities for men who wish to be seen as appropriately masculine (Connell, 1987, 1995; Seidler, 1991). Very few men can meet the narrow demands of this hegemonic masculinity – which offers up restrictions in terms of ability, age, class, race, sexuality, education, etc. Moreover, there are many women who occupy masculine subject positions (think about how women have to act "like men" in male-dominated industries or businesses just to survive).

In both theoretical and practical terms, it is important for men to challenge the essentialist categories of sex/gender that would otherwise silence us from contesting masculinism, sexism and patriarchy. We are inspired in this task by Gayatri Spivak who has a strategic response to the processes that serve to silence those who might otherwise contest sexism and racism. Her approach is set out in the following quote from an interview made more than a decade ago now:

> I will have in an undergraduate class, let's say, a young, white male student, politically correct, who will say: "I am only a bourgeois white male. I can't speak." In that situation – it's peculiar, because I am in the position of power and their teacher and, on the other hand I am not a bourgeois white male – I say to them, "why not develop a certain degree of rage against the history that has written such an abject script for you that you are silenced?" Then you begin to investigate what it is that silences you, rather than take this very determinist position – since my skin colour is this, since my sex is this, I cannot speak. (Spivak, 1990, p. 62)

The links between men and masculinity and women and femininity are thus neither automatic, nor essential. Instead, they are constituted in and through the social relations of sex/gender systems, and they must be seen as both relational (where identities are constituted in relation to other identities) and processual (where identity construction is a constant process of change). The following discussion, then, focuses on the hegemonic construction of masculinity. It is important to remember that this hegemonic subject position is neither totalizing (all-encompassing) nor monolithic (unchallenged). Indeed, hegemony can never be absolute nor uncontested, but is instead always partial and disputed. Further, as Gillian Rose (1993) argues, different men become masculine in different ways. It is useful to keep the above in mind, then, as we outline duppy feminism, a metaphorical way of thinking about and analyzing our contradictory and paradoxical participation in feminist geography.

Duppy Feminism

We think we want to be *duppy feminists*. Certainly we *want* to be feminists, or at least to conduct creditable feminist research. After all, as Michael Kimmel claims, "feminism provides both women *and men* with an extraordinarily powerful analytic prism through which to understand their lives, and a political and moral imperative to transform the unequal conditions of those relationships" (1998, pp. 60–1; emphasis in original). We also agree with David Kahane (1998, p. 224) that "at this historical moment, feminist research comprises some of the most interesting, creative and influential work" in the humanities and in the social sciences, including geography. But as Kahane also indicates, the road to male feminism is fraught with dangers, both for the personal and professional integrity of would-be male feminists and – more importantly – for the larger feminist project of developing structured analyses of patriarchy and practicing anti-patriarchal resistance and transformation. These dangers are partly about knowing (epistemology) and partly about doing (stance/practice).

In epistemological terms, how can men come to *know* and fully *understand* structures of patriarchal oppression that have victimized mainly women and benefited mainly men? (Or, to paraphrase Brian Pronger (1998), how can "bundles-of-desires-with-penises" experience "feminist desire"?) In terms of practice, what sorts of stances can pro-feminist men (who are always already embodied and socially positioned as male subjects) develop *vis-à-vis* feminism in order to introduce their own analyses, and perhaps their own subjectivities and experiences, into feminism without also introducing material and discursive effects that are detrimental to pro-feminist struggle?

In the essay "Male feminism as oxymoron," David Kahane (1998) describes four ideal–typical male feminist positions that do *not* successfully avoid these pitfalls of knowing and doing: the poseur, the insider, the humanist, and the self-flagellator. Although we suspect these ideal–types can be read in ways that imply an essentialist reading of men and male-ness, we feel if used cautiously and contingently, they provide useful categories for thinking about how many men articulate with feminism. Here we want to work from these types to describe some of the challenges of researching as male feminist academics, or perhaps more pointedly, to describe the range of epistemic violences available to pro-feminist male researchers. We will then propose a fifth alternative – duppy feminism – as a preferable, albeit an incomplete and partial way of imagining one's (male) self in relation to pro-feminist struggle and the more specific task of undertaking (male) feminist research. But first we have to take a detour on the road to a feminist male feminism, into terrain not often associated with feminism; a detour into the world of reggae, Rastafarianism, and "duppy business."

Rastafarians admonish one another to avoid "duppy business." This phrase can refer to a wide range of things, including Christian and revivalist beliefs in the holy spirit and life after death; widespread Jamaican folk preoccupations with evil spirits; and all forms of this-worldly deceit and duplicity – especially the neocolonial "politricks" employed against the "sufferers" (as Rastafarians describe themselves and other poor Africans living in Jamaica). Bob Marley and the Wailers draw upon the rich discourse of the duppy throughout their oeuvre, but especially on *Burnin'* (1973), their second album with Island Records, and most particularly in the classic reggae anthem "Duppy Conqueror" (see Dawes, 1999; Johnson-Hill, 1995). The main words of the song's chorus – "I'm a duppy conqueror, conqueror" – sung in dread, dirge-like cadence, sound initially like nothing more than a shallow rude boy boast/threat (the term "rude boy" describes variously a category of underemployed, impoverished Jamaican street youth, their characteristic personal style, and a type of music that emerged in the mid-1960s to express their experiences). But the Wailers are almost always subtler than that, and constantly aware of their responsibilities as reggae's most visible social and political theorists. In that context, these simple lyrics resolve quickly into a double play on words. Bob Marley, lead singer of the Wailers, is simultaneously a *duppy CON-QUEROR* and a *DUPPY conqueror*.

Marley's claim to be a "conqueror" slips nicely into the category of rude boy bravado (in one of the verses he sings "bars could not hold me; force could not control me, now"). However, if we understand "duppy" as describing the *nature* – rather than the object – of his conquering, we read something else. He is occupying an ephemeral role; a conqueror, a fighter, but in a ghostly, not quite visible, not quite tangible sense. He is a conqueror in a guerilla non-war. Such a conception is entirely commensurate with the Wailers' sense of their mission as reggae musicians, and also with a more general Rastafarian understanding of their resistance as being simultaneously unrelenting, intangible, and often disguised; a resistance of celebration, sufferation, remembrance, and selective non-participation (what Scott, 1990, calls everyday resistance and Foucault, 1980, describes as agonism; see also Thiele, 1990; Laclau and Mouffe, 1985).

But "duppy" can also be understood to describe the *object* of the Wailers' conquering. They are fighting duppies, specifically the duppies of a racist, colonialist past and present, the ghosts of slavery, hunger, political repression, etc. that haunt contemporary Jamaica, and which provide the themes for many Wailers' lyrics. In this sense "duppy" is used as a subtle and illuminating metaphor to characterize the ever-present, but often ephemeral, slippery and duplicitous forms of oppression against which the sufferers struggle.

The Wailers' deployment of the duppy discourse articulates a sophisticated form of resistance and claiming. But it also expresses a certain respect for the complexity of its object, Babylon System (one of the terms Rastafarians use as a catch-all for the range of discourses, institutions, agents, and practices of oppression operating against the sufferers) as well as an awareness of the limits – the necessarily ghost-like quality – of resistance. Both of these realizations relate to an awareness that the duppies against which the Wailers (and all sufferers) are struggling operate *within* and not just *on* the "conquerors." The Wailers (*especially* the Wailers as wealthy reggae superstars) are aware that they are constituted by and within the field of domination they are resisting; they are engaged in a struggle *against* Babylon System and *for* personal integrity.

For now, let us say that both spins on "duppy conqueror" provide lessons for pro-feminist men. First, perhaps a useful way to approach our own search for feminist knowledge (an epistemological issue) is as *DUPPY feminists*: always starting from the realization that we are fighting a duppy – patriarchy, male privilege – that is within us. This duppy helps constitute our subjectivities in duplicitous, often intangible but ever-present ways, and it commands us to find ways to use the inevitability of ambivalence and contradiction productively. That could be *our* standpoint. Second, maybe a useful way to approach doing (stance/practice) feminism, and feminist research is as *duppy FEMINISTS*: always there, unrelenting, insinuated into the discourses and institutions of patriarchy. But, as *duppy FEMINISTS* we are also ephemeral, in the background (e.g. in the acknowledgements rather than the references, or learning to work towards less authority rather than more) and strategically duplicitous towards the institutions of patriarchy (and not towards our feminist colleagues, as, for example, when they get jobs they are qualified for, but which we had assumed should be ours because of male privilege).

We realize that by constructing our discussion around the metaphor of duppy feminism we risk committing an act of cultural appropriation. This may not seem to be a promising foundation for an essay concerned with the problem of reducing the impact of certain types of structured privileges. However, the Wailers' use of the duppy metaphor is autoethnographic: that is, it is a deliberate effort by members of a subordinate group to express their reality in a language and idiom that members of a dominant group will understand, and that they will be able to relate to their own reality in some way (M. L. Pratt, 1992). We try to use "duppy conquering" in that spirit, as a sanctioned and respectful ethnographic appropriation – an evocation from and of a certain experience of marginality and ambivalence – which can help guide pro-feminist men through (but not out of) their own ambivalence and into a productive paradoxical space: a place from which to practice. Indeed, we think of our appropriation of the duppy metaphor (and its situation in Rastafarianism and reggae music) as a sort of (parallel) parable for what we mean by duppy feminism itself. We will return to – and further develop – this duppy/feminist analogy after a brief contextual discussion of Kahane's ideal types of the unsuccessful male feminist.

Kahane begins with a point similar to one we just made: "men have to face the extent to which fighting patriarchy means fighting themselves" (1998, p. 213). He engages with Sandra Harding's work on standpoint epistemology to accept cautiously that it is *possible* for men to develop productive and transformative feminist knowledge (see Harding, 1998). He is, however, more skeptical than Harding of the *probability* of such a development, which necessarily involves men in ongoing processes of internal struggle, often entangling them in convoluted self-rationalizations and "various forms of bad faith and self deception" (1998, p. 213). This is not an indictment of men, merely Kahane's insistence that it is difficult for us consistently and continuously to concede fully to the fact of male privilege in our own lives (and theories). It is even more difficult to work systematically against our own patriarchal interests (whatever we may *say* about patriarchy oppressing men and women alike) in attempting to dismantle structures and practices of male privilege. For example, it is one thing to acknowledge the reality of male privilege, quite another to attribute our own academic success as much to this privilege as to neutral merit, and another thing still to actively work to undermine the unfair advantage male privilege has given us. Even the most rigorously trained pro-feminist men among us may find the third (or second) stage in this scenario a bit much to expect: and there begins the process of "bad faith and self-deception."

Kahane observes that "bad faith and self-deception" are often most visible (among pro-feminist male *academics*, at least) in the space between theory and practice. Although we are suspicious of too strong a distinction between them (for example, is teaching feminist theory an example of practice or theory?), we take Kahane's point that "men have a lot to gain (and not much to lose) by dabbling in feminist theory, whereas they have little practically to gain from feminist practice, especially when this practice directly challenges patriarchal structures and behaviours" (1998, p. 224). This statement – and we think we can accept its general veracity – has interesting implications for men's attempts to do feminist social science research, because of the theoretical *and* practical (thinking *and* doing) nature of social science research procedures. Thinking our research with feminist thoughts without doing our research in anti-patriarchal deeds is – in part at least – an instance of bad-faith and a betrayal of the sort of guerilla/duppy feminism hinted at above. (At the same time, however, it is important to remember the pervasive character of masculinism and the difficulty that both men and women have in identifying it and contesting it.) In developing his ideal types of unsatisfactory male feminism Kahane is most concerned with the state of *being* a male feminist, but his types also allow us to interrogate some of the theory/practice pitfalls of *doing* feminist research as pro-feminist men. It is worth noting that these are *ideal* types, more useful for identifying tendencies than for definitively categorizing individual pro-feminist men, most of whom (ourselves included) exhibit characteristics of each of the types at different times and in different circumstances. Moreover, the "bad faith and self-deception" that Kahane speaks of is not necessarily a tactic employed by fully self-aware agents. Rather, because of the fractured and contradictory character of human subjectivity, many of these practices will be undertaken by agents acting through unconscious fantasies and desires.

The theory/practice divide noted above is most evident in what Kahane describes as *the poseur*. Here is an individual who is interested in feminist theory, has engaged in a pick-and-choose reading of it, and has used bits of feminist theory to serve his non-feminist projects. He has not, however, "turned feminist critiques upon his own theoretical and practical tendencies – to have extended his feminist analysis from a concern with structures to a consideration of how these structures have informed his own identity in ways that correlate with patriarchal harms" (Kahane, 1998, p. 224). This somewhat de-personalized commitment to feminism as purely propositional knowledge is attractive and comfortable for a couple of reasons. First, it allows men to feel good about their sense of community and allegiance with the "women's movement," without delving deep enough to feel bad about their own involvement in patriarchal oppression. Second, adopting the stance of *poseur* puts men in comfortable and self-affirming company, as many putatively feminist female academics also frequently occupy this position.

What kind of feminist research can we expect to emerge from the stance of *poseur*? This stance quite likely will lead to analyses of research findings that are informed by aspects of feminist theory. But it is unlikely that the processes of defining a research problem, developing methodologies, selecting and relating to participants and co-researchers, or collecting data will be much influenced by a will to practice anti-patriarchal resistance and transformation. This type of research can be useful for advertising the analytical power and flexibility of feminist theory (and the male feminist theorist), but does little towards feminist social transformation (or the development of feminist desire in the researcher). Indeed, this research scenario can be understood as a technology/resource of male privilege, as it transports male researchers into the lofty (and arguably prestigious) realms of feminist theorizing, while leaving the mundane task of transforming the research process according to pro-feminist principles up to women. In other words, this is feminist practice deferred by theory. The stance of *poseur*, in its success at deflecting feminist analysis away from the self, allows the male feminist to feel good about himself, while protecting him from the anguish of glimpsing his complicity in patriarchal oppression. A lot hangs on the balance of this delicate trick of deflection: turning to feminism to feel better about ourselves leaves us vulnerable to the eventual realization that "feminist analyses don't tend to cast men in a particularly flattering light" (Kahane, 1998, p. 225).

The insider resolves the paradox that the *poseur* is left with by combining his theoretical interest with a degree of political commitment to feminism: volunteer work, reading, pedagogy, and conversation. In this ideal type, these outward demonstrations of faith (which may be quite sincere) are not accompanied by sustained self-reflection. "The *insider* premises his comfortable self-image on doing well in the eyes of feminist women, even while his comfort militates against his addressing his own sexist tendencies" (Kahane, 1998, p. 225; emphasis added). Failure to turn critiques of patriarchy on himself, however, prevents the *insider* from recognizing the extent to which the appearance of "doing well in the eyes of feminist women" is itself premised on his secure position within a system of patriarchal privilege. He is unlikely to realize that he may be being humored by women because of his gendered power, his institutional influence or authority, his refreshing difference from non-feminist men, or merely because he so obviously expects to be appreciated. To employ Harding's (1998) distinction, he has developed a feminist perspective, but not a male feminist standpoint. The *insider* fails to recognize *himself* as an agent of patriarchy.

We can imagine that this failure would be key to the *insider's* approach to research. Certainly, his analysis will be guided by feminist theory. And, indeed, the processes of defining a research problem, selecting participants and co-researchers, and collecting data may be influenced by feminist principles. But the research process is unlikely to be characterized by any sustained effort to relate to participants and co-researchers in anti-patriarchal ways, or to employ methodologies for producing knowledge that relinquish cognitive authority to the women involved in the research. The *insider* remains the privileged male captain of a feminist ship. The research that results may contribute creditably to a body of feminist academic knowledge (developing structured analyses of patriarchy), but again, will do little practically to help dismantle male privilege. Indeed, the *insider* uses his feminism to appropriate and expand credit, power and credibility – arguably a way of increasing rather than dismantling his own male privilege, while avoiding the contradictions inherent in his position.

Many men read feminist theory deeply enough – or are sufficiently aware of their social surroundings – to get glimpses of certain patriarchal aspects of their own identities and social positioning. This can be an important step towards developing a male feminist standpoint, although positive results are not guaranteed. *The humanist* has experienced glimpses – perhaps full panoramic views – of his imbrication in patriarchy, and deals with the accompanying anguish and existential discomfort by concentrating less on the ways he has benefited from patriarchy than on his experience of its constraints. The *humanist* is interested in the various ways that patriarchy hurts men as well as women (e.g. by constructing men as competitive, casting them perpetually in the provider role, stunting their emotional development, or alienating them from their children). Such a preoccupation often involves a deep and sustained engagement with feminist theories of patriarchy, and a sincere attempt to become a different kind of man, "more in touch with feminine qualities and less constrained by patriarchal social structures" (Kahane, 1998, p. 227). While this is a worthwhile endeavor that may, in certain circumstances, improve the lives of some women, its primary concern is with the well-being of men. Feminists note that this is not a new preoccupation for men, nor a particularly efficient way to develop better understandings of the ways patriarchy systematically benefits men and harms women.

What sort of research can we expect to emerge from a *humanist* stance? Again, feminist theory – or the influence of feminist theory – will be prominent, and participatory/conversational methods often associated with feminist research are likely to be employed. But, given a primary interest in men's experiences of patriarchy, neither the selection of participants and co-researchers, nor the research methodology (which is likely epistemologically to privilege expressions of male knowledge and interpretation), are likely to do much to increase women's power and cognitive authority in the research process, nor introduce practices that decrease the effects of male privilege on *women's* lives. This is not *necessarily* the case (for example, it may be fruitful to deploy a full range of feminist research principles to the task of asking female participants to discuss the effects of patriarchal norms on their male partners' emotional development). But for the most part, *humanist* male feminist research will involve men using selected feminist theory to talk about themselves (again). This is unlikely to have a transformative influence on women's lives. Additionally, the discursive effects of men talking about their own beleaguered subjectivities from within feminism (whatever it is they say) might be detrimental for pro-feminist struggle.

Kahane's fourth ideal type is *the self-flagellator*. Men, when they adopt this stance, accept feminist analyses of patriarchy and male privilege, and recognize their own imbrication and complicity in these processes. Not a bad start, except that the *self-flagellator* deals with this recognition by focusing disproportionately on his own crisis of guilt and absolution. His objective becomes purging himself of sexism, and thus finally living up to "the feminist challenge" (Kahane, 1998, p. 227). There are two main problems with this stance. First, it is intolerant of – or perhaps blind to – the ambiguity, contradiction, and ambivalence that necessarily constitute male feminist endeavors. Even as men strive to constitute ourselves as part of the "solution" to patriarchy, we must realize that our socialization as men in a patriarchal society continues to constitute us as part of the problem. Failure to accept this contradiction – this structured failure – encourages frustration, self-deception, and perhaps eventually a "retreat to the stances of humanist or insider, or from feminism altogether" (Kahane, 1998, p. 227). Second, the will to meet "the feminist challenge" can itself manifest a perverse sort of competitive masculinism, a desire to regain control of the ball, a sort of petulant retort to feminist criticism.

The stance of *self-flagellator* does suggest certain avenues for research. Feminist theory will be prominent, as will conversational/interactional methods (especially biography, life history, and autobiography) commonly associated with feminism. However, to the extent that research is likely to focus on men's struggles to overcome their own imbrication in patriarchy, the research process will itself do little to resist male privilege or transform women's lives. Its practical/processual effects are likely to be similar to *humanist* male feminist research, although it may be more effective in stimulating self-reflexivity and articulating strategies men have used to develop themselves as less patriarchal subjects.

We can discern a number of irregular regularities among the research implications we have sketched for these four stances. First, all encourage the use of feminist theory, although to varying degrees. Second, while they differ in the extent to which feminist considerations are likely to influence selection of topic, methods, co-investigators, and participants, none of them require – or even really encourage – that the male researcher relinquish cognitive authority or male privilege in the course of the research process. Third, none of the four stances advocates research on topics that are likely to be particularly relevant to women's struggles to overcome patriarchal oppression, and two of them actively militate against the use of female research participants (except through the use of considerable imagination in designing the research). In short, each of these male feminist stances seems to encourage the use and further development of feminist theory by men, but not the relinquishing of male power, privilege, and cognitive authority in the day-to-day practice of putatively feminist research. It is worth noting that this is not just an indication of the personal venality of the various sub-species of male feminist, but also a manifestation of the operation of deep masculinism in institutional academic culture. We thrive in the university – sometimes women and almost always men (because of male privilege), sometimes feminists and almost always non-feminists (because of the anti-feminism and male privilege of the academy) – by appropriating theory and protecting personal power, control, and authorship. Pronger describes this deep masculinism (and this masculine desire) in terms of the territorialization of space; the will to penetrate others' space and protect our own:

This is the particular masculine desire for the territorialisation of space: it is the desire to conquer and protectively enclose space, the desire to make connections according to the laws of spatial domination. Here the capacity to exist is circumscribed by the will to control space and the fear of the violability of the same. It is a fetishised neurotic form of desire that appreciates its own existence in so far as it is in control of its space. Loss of control of space is the death of masculinity. (Pronger, 1998, p. 72)

Each of the stances described by Kahane exhibits strong traces of this territorial will to expand our territory. In this sense, feminist theory is used to penetrate the terrain of feminism; but the gates to our own territory, the boundaries for accessing the terrain of male privilege (not as an internal struggle, but as a social practice) remain tightly closed. How can men learn to treat our feminist research endeavors as something other than expansionary expeditions?

Toward Practicing Duppy Feminism

Maybe we can imagine the metaphor of *duppy conquering* as providing a preferable alternative to the metaphor of territorial conquering employed by Pronger to describe masculine desire. Maybe the exercise of imagining ourselves as *duppy feminists* can help us develop a less expansionary involvement in feminist research.

The first step is to recognize the duppy-character of patriarchy. Duppies, like other ghostly phenomena, find their homes in the uncertain spaces between consciousness, social relations, and material social practices. In societies with a strong duppy-sensibility, duppies are ever present, sometimes with tangible and material effects, but always as a threatening potential, a possibility waiting for an opportunity, a constant gnawing constituent of subjectivity, but not fully under the control of the subject. Marley's claim "I am a duppy conqueror" articulates a decision to reject the duppy of Babylon System, but in the interpretation we offered above, it also implies a recognition that this duppy is a part of him, but not fully part of his self-consciousness. Similarly, men's commitment to feminism must include the recognition that we cannot hope to be fully conscious of the ghosts of our own masculinity or those of patriarchy in general. We must learn to work productively with the knowledge that these ghosts are part of us, manifest in our (sub)consciousness, social relations, and material social practices; and that they must be engaged at each of these sites simultaneously, and without hope of definitive outcome. To a large extent this must be a recognition of the inherent contradictions of our position: struggling against male privilege from a position of male privilege that we cannot abdicate fully. Thus arises the paradoxical question: how can we use male privilege against itself in our research endeavors?

The second step may be to learn something from duppy tactics, as the Wailers suggest that Rastafarians have done. If Kahane's ideal types are any indication, there are strong temptations for male feminists to gather mainly, if somewhat unreliably, around the most visible, remunerative, respectable and institutionalized area of feminist scholarship: feminist theory. This has worked well for male feminists, but has not contributed significantly to dismantling male privilege, or to the project of using male privilege against itself. Duppy tactics would suggest something closer to a continuous vigilant opportunism; a preference for peripheral sites of intervention, away from the centers of socially and institutionally sanctioned activity; an attentiveness to micro-practices that run against the grain of patriarchal structures of merit, credit, and authorship; a willingness to exploit the inconsistencies, ruptures, neuroses, and contradictions of the patriarchal discourses that constitute us. This is a guerilla feminism for the academy. In practical research terms, it may mean relying less on developing theoretical credentials to justify our work as feminist, and concentrating more on exploiting the innumerable specific and contingent, and only partly predictable, opportunities to relinquish masculine territory – to spend male privilege. Spending such male privilege should take place in the definition of problems, the selection of co-researchers, participants and sites, the development of research relationships and data gathering procedures, the ways questions are asked, the ways analyses develop – the entire range of research practices. In even more practical terms, at each of these stages of research, it may reside often simply in the decision to listen rather than speak, to be impressed rather than impress, to give credit rather than claim it.

The third step is to choose these practical tactical maneuvers with a sensitivity about their (metaphorical) territorial implications. Male privilege is a territory. Patriarchy describes (among other things) the resources and technologies for expanding and protecting that territory. Pro-feminist men cannot desert that territory; we cannot abdicate male privilege. But perhaps we can find openings, routes into the terrain of male privilege, and opportunities for forays beyond its borders. Perhaps this greater boundary porosity will help dilute male privilege, allow the wider distribution of a less "potent" formula. Without suggesting for a moment that pro-feminist men should lessen their engagement with feminist theory, we nevertheless doubt that purely theoretical endeavors will pry open those gaps in the boundaries of male privilege. Pro-feminist male researchers would do well to seek these openings – these opportunities for dilution – along the meandering route we forge through the social relations and material social practices of the research process. That, we think, might be duppy feminism.

ACKNOWLEDGEMENTS

Authorship is listed according to the alphabetical order of our first names in order to highlight the "intimacy" of co-authorship, as well as to de-emphasize the masculine cultural ritual of lineage by which men's last names denote ownership. David would like to thank Kathryn Besio, Nancy Cook, Tania Dolphin, and Pamela Moss for their tolerance of his individual acts of bad faith and self-deception, and also Paul Berkowitz for suggesting the term "duppy feminist." Lawrence would like to thank many feminist colleagues – but especially Robyn Longhurst, Wanda Ollis, Karen Morin, Pamela Moss, and Kathleen Gabelmann – who have helped him learn about feminist theory and practices, all the while recognizing and accepting the contradictory spaces of feminism that he inhabits so uneasily. Although he is not always gracious about it, Lawrence is nevertheless quite grateful when his colleagues *refuse* to accept his acts of bad faith. We are both grateful to Pamela Moss for her constructive comments on earlier drafts of this essay.

RESEARCH TIP

Collaborative Projects

- A collaborative project will benefit from a clear delineation of group goals, expectations, and individual responsibilities from the outset. You may wish to draw up a formal agreement or 'contract' that outlines the tasks set for each member, making sure to establish deadlines.
- Set up a regular schedule for group meetings, including meetings with a supervisory member if appropriate.
- Work with each member's strengths, giving credit for a diverse range of contributions.
- Acknowledge tension and identify outlets for dealing with frustration and identify strategies for assisting the group process.
- Ensure that the size and scope of your project is manageable; group projects generally take longer than anticipated.
- Allow plenty of time for analysis, editing, and proofreading.

6

Toward a More Fully Reflexive
Feminist Geography

Karen Falconer Al-Hindi and
Hope Kawabata

Feminist researchers collecting the life stories of Maori women engage in *whakawhanaungatanga*, a Maori concept which refers to the process of building family, or kinship (Napia et al., 1999). As an offering of good will, trust, and reciprocity, the researchers tell their own life stories to the women whose stories they wish to hear. *Whakawhanaungatanga* helps to bridge the difference between the researchers and the participants, and establishes a permanent bond between them. Offering their own stories as gifts prior to receiving the gift of another's story places the researchers on a par with the participant. Such an approach to interviews would be both impractical and culturally inappropriate in most of the English-speaking academy. Yet, we can apply the ideas of shared vulnerability and comparable positioning for researchers and participants, and this is one of the goals of the approach that we advance in this chapter.

As Kobayashi states, "difference – or its putative opposite, sameness – is an inevitable ontological condition that is never completely achieved" (1997, p. 3). Feminism grew out of a concern with difference, and geographers who recognized that women are often viewed as different from men (and not only as different but as *inferior*) felt the need for a specifically *feminist* geography. Learning about difference and presenting this knowledge in a meaningful form is part of social science research; at the same time, interpersonal difference permeates the research process. For example, this chapter grew out of a collaboration between Hope and Karen that began with Hope's MA thesis; the illustration later in the chapter comes directly from Hope's research. As Hope designed and conducted her study, and later wrote about it, she grappled with issues that arose from differences between herself and her research participants. Hope and Karen (Hope's thesis supervisor) also experienced differences between them over the course of Hope's research and during the writing of this chapter.

Because this chapter concerns difference and research methodology, we want to expose rather than conceal our individuality and disagreements. An earlier draft used the "we" commonly found in co-authored texts, but Hope objected on the grounds that Karen had developed the arguments well beyond those Hope made in her thesis. Karen and Hope agreed that the chapter should be written in the first person singular, but that person is Karen in all sections but the fourth, in which that person is Hope.

That we are concerned about such issues and can articulate our compromises reflects a contemporary, ongoing, transformation of social science. Ruth Behar, for example, writes about the search for new ways of researching and writing anthropology that will connect the scholar and the researched and which emphasize the humanity and frailty of each (Behar, 1996, p. 5). Feminism's commitment to understanding the world in order to change it (Acker et al., 1993) requires that feminists work toward a methodological strategy that helps scholars as well as others to understand rather than obscure difference, values participants' knowledge as well as the researcher's, and shares power rather than reinforcing hierarchy. Reflexivity has been identified in the literature as one of the ingredients of this revolution. In this chapter we argue for a more "full" version of reflexivity that can make a strong contribution to the transformation that is at hand. Feminist geographers are ideally placed to take up the promise of a fuller reflexivity and to employ it in their own transformative work.

As understood in everyday language and typically used in academic discourse as well, *reflexivity* means *reflectivity*, or the act of reflecting upon oneself and one's experiences (see for example England, 1994). To researchers, this usually means "giving as full and honest an account of the research process as possible, in particular explicating the position of the researcher in relation to the researched" (Reay, 1996, p. 443). Many feminist geographers struggle with such an idea of reflexivity. In this chapter I "push" the concept, and argue that a more fully reflexive research practice can be useful. Because I conduct research using a semi-structured, in-depth interviewing method of data collection, I use interviews within a feminist geography research framework as example. I wish to make the argument, however, that the fuller reflexivity I advocate may be broadly applied.

Following this introduction, the chapter continues with a stage-setting overview of feminist research and research methods. That section concludes with a discussion of reflexivity in feminist geography research as it is conventionally practiced. The next section offers an argument and model for an extended reflexivity. Then, Hope provides an illustration of such an approach in practice. Finally, I link the argument to others' calls for methods that de-center the researcher and that encourage her to relinquish – some! – control over the research process.

Toward Reflexive Feminist Research

Feminist social science research is explicitly political. It is distinguished from mainstream social science, and from other alternative approaches, by its concern to reveal and, often, to redress gender inequality. Consistent with this agenda, feminists seek to empower individuals and groups that have historically lacked access to power; this includes women, of course, but also racial and ethnic minorities, working-class men and women, those with physical disabilities, and others whose difference from a white, male, middle-class, able-bodied and heterosexual model of normality have disadvantaged them. Middle-class, white academic feminists are aware, spurred in part by a trenchant critique launched by women of color (see for example Lorde, 1984; Spivak, 1988; hooks, 1992), that they enjoy a privileged position. Many seek to redress this through their research practices. But how can such scholarship be unbiased and valid, given its avowedly political stance?

All social scientists, feminist and otherwise, place a premium on valid research. Validity refers to the trueness of an account or measure: "we want to know that our research results fairly and accurately reflect the aspects of social life that we claim they represent" (Acker et al., 1993, p. 145; see also Bailey et al., 1999). Bias is a threat to the validity of a study and may enter the research during data collection, data analysis, or interpretation. Guards against bias are part of every methodological strategy. In the case of semi-structured interviews, for instance, a participant may speak at length to answer a question, and the researcher usually analyzes such an answer in its entirety and reproduces it whole in written accounts. Close attention to a detailed answer from a participant is one guard against a poorly worded question, an unfair analysis, or a skewed interpretation. A valid research-based claim, therefore, is unbiased, and may be described as "objective."

To many feminist scholars, a political approach to research stands a better chance of producing valid results than an approach, such as positivism, that claims to ignore power relations. Sandra Harding, an advocate of feminist standpoint theory, and Donna Haraway, whose explication of situated knowledge has become well-known among geographers, have articulated such approaches. According to standpoint theory, certain groups of knowers are better positioned, because of their disadvantaged social situation, to analyze the processes through which they are disadvantaged. Standpoint theorists advocate "studying up" by looking up through, for instance, the "glass ceiling" on women's advancement in corporations in order to understand this barrier with greater clarity than is possible if one "studies down," from the point of view of those who benefit from sexism. Rather than a static position such as that suggested by the notion of a standpoint, the idea of situated knowledge views individuals as multiply positioned within different frameworks of power, race, class and gender, and calls for knowers to take responsibility for what they claim to know with respect to their positions. Such understandings of where knowledge comes from inform the ways feminist researchers approach different methods of research.

Increasing openness to qualitative research methods combined with growing skepticism toward standardized data collection tools (and the quantitative analyses these facilitate) mean that interviews, especially semi-structured and open-ended ones, are gaining in popularity throughout the social sciences. The important difference between this approach and the more traditional, structured interview is that the path of the conversation between researcher and participant is not predetermined, nor is the spontaneity inherent in the flow of conversation truncated. Despite its popularity and increasingly sophisticated use, a problem continues to bedevil the open-ended interview – expressed as "difference between interviewer and interviewee" (see for example Reay, 1996; Song and Parker, 1995); people wish to learn from and about others because the latter are different from the former, but the *fact* of difference itself may distance them from one another, making such understanding difficult.

Mainstream social science research has always had difficulty with dissimilarities between interviewers or researchers and research participants or interviewees. Specifically, differences between researchers and participants in age, life stage, race and ethnicity, class, gender, (dis)ability, and so on, are believed to create barriers to rapport during the interview, in which the interviewee (or, more often, "informant"), is the source of valuable data to be disclosed to the interviewer (see figure 6.1(a)). Many mainstream interviewers try to minimize the influence of the interviewer on the participant: the interviewer is a *"neutral* medium through which questions and answers are transmitted" (Babbie, 1973, p. 172; emphasis in original). Many feminist researchers as well as others, of course, have disavowed this purportedly neutral stance. Instead, the interviewer (more often, the researcher) is certainly an *instrument* of research but is anything but a neutral one (England, 1994). Both power and difference, or rather *identities*, are at play in the field of relations in which researcher and participant are located.

Against the mainstream approach to interviews, the feminist researcher engages her participants in open-ended, semi-structured, interactive interviews or *conversations* in which the nature of the interaction itself produces new information or insight about a topic (see figure 6.1(b)). Initially, many feminists assumed that because women interviewed women, they had so much in common *as women* (the experiences of childbirth, mothering, and

Figure 6.1 Models of feminist reflexivities

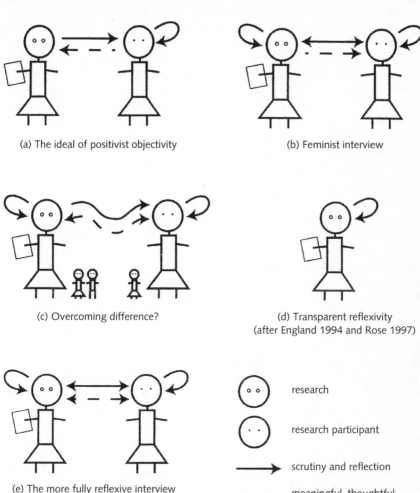

(a) The ideal of positivist objectivity

(b) Feminist interview

(c) Overcoming difference?

(d) Transparent reflexivity
(after England 1994 and Rose 1997)

(e) The more fully reflexive interview

○ ○ research

(‥) research participant

⟶ scrutiny and reflection

– → meaningful, thoughtful
communication

sexism, for example) that other differences were unimportant. White feminists in particular were shaken out of this essentialist position by African-American and other minority scholars and activists who argued that gender was not the only dissimilarity that mattered, either in life or research (see for example hooks, 1990). A more nuanced appreciation of variety among women is one outcome of the feminist postmodern and poststructuralist critique of positivism. Many researchers, however, continue to be frustrated by research methods which are relatively clumsy by comparison to the sophistication of feminist epistemology.

Feminist interviewers employ various strategies as they seek to establish commonality across difference with their participants. Gibson-Graham (1994), for example, asked her research associates to interview their friends in order to take advantage of their pre-existing trust, while Oberhauser (1997) took her children with her to interviews so as to build rapport with her Appalachian homeworker participants through their shared interests as mothers (see figure 6.1(c)). By incorporating their roles as friends and/or mothers into the interview process, these researchers make themselves vulnerable to their participants in parallel to the exposure of the participants' lives in the interview. This helps to "level the field" between researcher and researched and so disrupts the traditional hierarchy which places the researcher "above" her participant. It also introduces the heterogeneity of the researcher's life into the data-collection situation, so that she becomes multi-dimensional in the same way that many feminists want to represent the lives of their research participants: complicated, compromise-ridden, and sometimes contradictory.

Gillian Rose (1997) identifies the researcher as the "knowing analyst" of the researcher–participant landscape, and criticizes feminist geographers' conceptualization of difference-as-distance for reproducing the ostensibly all-knowing, objective vision of the positivist scientist. But Rose's criticism is not entirely fair. That is, while the "landscape of power" of feminist research may *appear* to replicate the positivists' claim to all-knowing objectivity, the two approaches are in fact quite different (compare figures 6.1(a) and (b)). Of principal interest is the fact that traditional interviews, consistent with a positivist epistemology, require that the respondent be reflexive about her/himself and share her/his insights with the interviewer. Most interviews conducted within a feminist research agenda also require this participant reflexivity, but – and here is the important difference – require in addition that the researcher *reflect back to the participant* the researcher's understanding of the participant's thoughts. One benefit of this practice is that it affords the participant an opportunity to correct any misunderstanding between them. Further, this approach encourages self-reflexivity by the researcher, although insights gained thereby are not usually shared with the research participant. In other words, one expects the researcher to pose additional questions to the participant based on the participant's reflective statements, but the researcher does not usually "think out loud" to the participant about the researcher's self-evaluation *vis-à-vis* the research. Thus, Rose fails to appreciate the advance that feminism offers over positivism in its approach to interviews. An extended notion of reflexivity further advances the methodology of interviews conducted within a feminist research epistemology.

Extending the Notion of Reflexivity

Where the identities of the researcher and research participant are discordant, reflexivity is advocated as an aid in pushing past this barrier (Hertz, 1997). Thoughtful reflection on one's research practice, one's subjectivity relative to that practice, and self-criticism and change where warranted would certainly seem to improve the process and outcome of methodologies in which the researcher herself is an instrument of the research (see figure 6.1(d)). The limits of this conception of reflexivity are demonstrated by the presentation of reflexive practice in scholarly writing. Reflexive researchers take one or both of two approaches when including their reflexive stance(s) in research reports. One is to share introspective notes from a personal or field journal. For example, Diane Reay wrote of her anxiety about her subjectivity and its limitations for her fieldwork in her diary (Reay, 1996, p. 446). This tack permits the author to reflect further about the intersubjective nature of her work, as well as make clear to readers that her practice has been reflexive. A second approach is simply to describe one's circumstances so that readers may ostensibly decide for themselves what impact the individual's life and self has had upon the study. Certainly both representations of a (conventionally) reflexive approach are an improvement over the alternative, implicit assumption that the researcher is an "objective" reporter of the research. Unfortunately, neither offers much in the way of advancing a study. Instead, they may aid in the evaluation of a study, or just help the ethnographer learn about her own life (Wasserfall, 1993). Such a researcher-focused use of reflexivity seems to truncate the possibilities inherent in a powerful idea. That is, reflexivity here is limited to a solitary consideration of oneself, as though one were gazing at a single reflection in a mirror and ignoring the myriad reflections from the other mirrors lining the walls of the room.

Kim England (1994) and Rose (1997), for example, have each written of "a sense of failure" regarding their attempts to conduct interview-based research consistent with feminist, post-positivist principles. England, for instance, did not pursue a particular line of research inquiry because of issues of racialized and sexualized identity, and power, concerning her research assistant. And, although Rose completed her project, her inability to interpret and understand an interviewee's joke about their respective nationalities continues to trouble her. In each case, self-reflexive introspection has been unable to resolve the troubling issues raised by the feminist alternative to positivist inquiry.

Ethnographers, in particular, have begun to push the idea of reflexivity beyond application to the researcher, by the researcher, to include research participants' insights. In her participant observation of gender, hierarchy and space in West Africa, feminist geographer Heidi Nast extended the idea of reflexivity to include "learning to recognize *others' constructions of us* through their initiatives, spaces, bodies, judgment, prescriptions, proscriptions, and so on" (Nast, 1998, p. 94; emphasis added). Nast had several uncomfortable experiences during her fieldwork in Nigeria (see Nast, 1998, for a more complete discussion). Each concerned Nast's body: its placement, dress, or movement. For example, shortly after her arrival in the city of Kano, Nigeria, Nast decided to visit the royal palace. Her preparation included donning an elaborate head covering and garments she believed were appropriate for the visit. Her knowledge of Hausa language was adequate; however, to understand the remark made by one concubine to another as Nast entered the palace concerning the "idiot" passing by. Later she learned that she had not covered her head properly for someone of her social situation, and that this was the occasion for the concubine's remarks. This as well as other incidents helped Nast to learn about the person or people who *placed* her based on their interpretations of her: "Feeling their surveillance and listening to their sexualized comments helped me recognize *them* as much as they helped me recognize myself in terms other than my own" (Nast, 1998, p. 109). Similarly, anthropologist Lila Abu-Lughod has described feeling utterly naked before a Bedouin elder because her head was uncovered, not because that is how her culture defined nakedness, but because it was how *his* did so (Abu-Lughod, 1988). This conceptualization of reflexivity points toward an understanding of the constitution of difference, or identity, in the research process (Rose, 1997, p. 313).

Nast's and Abu-Lughod's operationalizations of reflexivity extend the meaning and utility of the concept much farther than usual. Instead of reflexivity as *reflectivity*, or gazing at oneself, these scholars are centrally interested in themselves only in the gaze of others (see figure 6.1(e)). And, far from being satisfied to learn about themselves through others' eyes, they *return* the look – now informed by the "others'" gaze – to the other in order to learn from her: how she (the participant) sees/places/interprets me (the researcher) tell me about her, her worldview, her categories? By redefining the use of oneself as an instrument of research, these scholars are positioned both to learn more about the phenomenon under study *and* to share power with their research participants.

The more fully reflexive approach is readily employed in ethnography, or participant observation, as we have seen. The sustained engagement between researcher and subjects required by the participant observation method provides ample opportunity for encounters such as Nast and Abu-Lughod describe to occur, and for the researcher to reflect upon and learn from them (see also Nagar, 1997). But in-depth, semi-structured interviews can also be more fully reflexive. For instance, in their studies of cultural

identity among Chinese youth in Britain, Song and Parker (1995) emphasize that their participants positioned their own identities relative to those
they presumed were the authors'. Although neither author intended to do
so, over the course of their research projects they found that they had *de
facto* used their 'selves' – their bodies, their names, their identities – as
integral components of the research.

These bodies/selves must be found some*where*, of course, and geographic
space is central to a more fully reflexive approach in feminist geography.
In the following illustration for example, I highlight the role that leaving
Ou's office for the elevator lobby seemed to play in shifting the tenor and
content of her conversation with Hope. Although the relational space that
the two of them had established during their interview did not change,
their physical location did, and with it came new possibilities for their
interaction. Similarly, in Nast's retelling of her fieldwork experiences in
Kano, Nigeria, social space plays a prominent role. She became the object
of critical, sexualized remarks when she walked through the concubines'
area of the Kano Palace, and came to an understanding of their view of her
after she understood their social and physical *place* within the social
hierarchy. As Soja (1980) and Pred (1984) remind us, social and spatial
relations are dialectical and ever-becoming. Women and men transform
places through their interactions in and with them; simultaneously these
gendered beings are defined and perhaps altered by the places themselves.

If the researcher is open to self-critically reflexive involvement with her
participant (as in those instances described above), an interview methodology can incorporate a more full reflexivity. In fact, I suggest that England's
(1994) and Rose's (1997) reported "sense of failure" regarding their
respective research projects might have been mitigated by such an
approach. Drawing on Foucault's synaptic, productive conception of
power, Julie-Kathie Gibson-Graham (1994) and Rose (1997), for example,
recognize participants' knowledges as powerful because they are productive. If feminist researchers really believe in sharing power and validating
the knowledges of research participants, we must pursue their perspectives
and invite their observations, no matter how uncomfortable for us these
may be. I have found such an approach very illuminating, as Hope shows
in the following section.

Hope's Illustration

I (Hope) undertook in-depth, semi-structured interviews with ethnic minority women who worked from home, in 1997 (Kawabata, 1997). The
purpose of my research was to investigate the mutually constitutive relations among geography (in particular, spaces of the home), gender, ethnic-

ity, and work: did performing regular paid work tasks from home using telecommunications technology such as a computer modem rather than physically commuting to the workplace reinforce or challenge these women's ideas of themselves as members of specific ethnic groups, as paid professionals, and as women/mothers/wives?

When I (I am Japanese) asked Ou Johnson (pseudonym for an American woman of Chinese descent) about how women's roles in Chinese culture are different from American women's roles, Ou answered "I think they are similar. I think in your culture, I can see it being very different." I then asked, "do you mean Japanese and American are different, but Chinese and American are kind of similar?" She answered "yes," based on her perception of her own mother, whom, Ou believed, acted like an Anglo woman. She also based her answer on her mother's view of Japanese women, whom Ou's mother viewed as totally different from Chinese women.

Ou perceived me positively, as an Asian, and as a foreigner. Ou has never been friends with Chinese Americans, Chinese people, or even other Asians, in part because there was only one other Chinese family in the midwestern city where she grew up. She mentioned to me the discrimination she faced during a visit to California from Chinese Americans there. She did not feel comfortable with Chinese Americans, but felt comfortable with me. She told me that if I were part of the Asian group, she would be willing to try to get to know them. So Ou saw me as a foreigner and as an Asian, in contrast to herself, whom she viewed as an "American Asian." She roughly compared her identity with what she imagined mine to be: "I feel like I'm American ... with the Chinese background, versus you, Japanese, just the opposite" (Kawabata, 1997, p. 56).

Ou wanted to know my Japanese or Asian point of view on certain matters. This was why, following the conclusion of our "interview," Ou walked me out to the elevator of her office building, where our conversation continued for some minutes. I was taken aback when Ou asked me: "How about you, Hope? I bet you have been discriminated against with your *nationality* in this country, haven't you?" After I discussed this interaction with Karen (and after I had dealt with my shock), I came to an interpretation of what Ou meant. I believe that I was surprised that an Asian woman would ask another Asian woman this question. Karen suggested that Ou, perhaps with a lingering Chinese prejudice against Japanese people and with knowledge of the Japanese internment during WWII in this country, assumed that I would face discrimination by Anglos. Karen further suggested that Ou's question shocked me because of a remnant of prejudice against the Chinese on my part. I do not agree with this at all. In any case, I understand with stark clarity how different from myself and how like Anglo-Americans Ou perceived she (Ou) was.

Karen's Interpretation

Both physical and relational space are crucial in the process of the more fully reflexive interview. Here, by walking together out of Ou's office and to the elevator lobby of her building, Ou and Hope marked the end of the formal interview and signaled a transition to their meeting's conclusion. Having left the space of her office, Ou may have felt that the focus was no longer on her, and perhaps felt more at liberty to remark on her interpretation of Hope's identity and situation as a foreigner in the US. Similarly, Song and Parker (1995) interviewed young Asian immigrants in their families' places of business. Both researchers found themselves identified and categorized in sometimes uncomfortable ways that revealed much about their participants' views of the world and themselves. The researchers' identities and selves may have been more "up for grabs" on the interviewee's "turf."

These examples highlight the power of a fuller reflexivity in feminist interviewing. Through the view of herself in the eyes of the (O)ther, Hope formed an interpretation of herself (how she might be perceived by some other Asians, especially those of Chinese heritage) and, more important from a research perspective, about Ou's placement of herself in relation to Hope. This latter is the important contribution of the reflexive stance to feminist research. Hope learned about *how different* from "Asians" and "foreigners" Ou viewed herself, despite the fact that others (Anglos, say) might see her as very different from themselves. Ou's relational "placement" of Hope, for example, told her a great deal about Ou's understanding of her *own* position. Ou does not see herself as an outsider to an American nationality; in fact, she claims such an identity as her own. Regarding the research, this means that the fulfilment or avoidance of an ethnic/racialized minority role could not have motivated Ou to pursue a telecommuting work arrangement. Drawing upon her interpretation of Hope's identity permitted Ou to communicate her view of herself more clearly than she did at any other point in the interview.

Possibilities

What can a researcher do to ensure that her interviews are more fully reflexive? Most importantly, after Nast (1998), I recognize that true reflexivity means letting go, allowing the possibility for being out of control. Nast is interested in the out-of-controlness of situations where our bodies are inscribed, prohibited, or disciplined according to others' worldviews and interpretations of our bodies; I am more interested in the loss of control that interviewers can experience in a truly non-hierarchical conversation with a research participant, in which that participant feels comfortable enough to reveal to me her positioning of me relative to herself or her experiences. Hope's experience with Ou illustrates the unexpected nature of others' responses to and positionings of the researcher. Willingness to give up control over the interview, as well as to share information about oneself, are keys to the possibility of a more full reflexivity. As Hanson has written: "I would like to see us devise methods and methodologies that maximize the chance that we will see things we were not expecting to see, that leave us open to surprise, that do not foreclose the unexpected. . . . Pursuing this goal is likely to entail a letting go, a conscious attempt to relinquish control over the research process" (Hanson, 1997, p. 125).

As social science research, and feminist research specifically, transforms itself from an endeavor that attempts to follow a "hard" science model to a project that writes about people/ourselves and the world as people/we experience it, scholarly writing is irrevocably altered, as well. Writing about research conducted in the more fully reflexive mode that I have described here requires that the researcher identify and locate herself, not just in the research, but also in the writing. She must be willing to write and so re-live discomforting experiences, to look awkward and feel ill at ease. She must commit to paper and thus to the scrutiny of peers and others that which she might prefer to forget. But the pay off, I feel, will be worth it.

The more fully reflexive approach offers three advantages. First, rather than targeting difference *per se*, a fuller reflexivity helps one to understand how identity is constituted during the research process itself. I want to know how the research participant locates herself within the world as she lives it, and I recognize that identity shifts in relation to context; the research situation and researcher provide a context against which the participant may "bounce" her sense of self. Second, by providing an additional, positioned view of the researcher, a fuller reflexivity helps to make the researcher's positionality *vis-à-vis* the research more clear. For example, Hope's conversation with Ou revealed the possibility of an unexpectedly strong ethnic nationalism. Finally, this approach may help to share power with the participant more equitably than is possible with other methodologies, and validates the participant more fully as a knower. That Ou could pose the question she did to Hope suggests that Ou was empowered by the interview process and so made a more full partner in it. More fully reflexive research, and writing, bring the scholarly community as well as the larger communities of which each of us is a part more and better information about the world(s) which we seek to understand.

ACKNOWLEDGEMENTS

Hope would like to thank Karen, who was her thesis advisor. Karen helped Hope think critically on her journey to Christian feminism. Hope dedicates all her graduate work to her grandmother Kitazawa who passed away in 1996. Above all, she returns all the glory to the One who created her. Karen would like to thank Hope for her collaboration on this project. She is grateful for the insightful suggestions and warmth of The Writing Group, whose members "aspire to wise." She dedicates this chapter to her son Sean, whose gestation and first year coincided with its writing.

RESEARCH TIP

Jargon in Feminist Geography

What can be particularly daunting for the initiate of feminist geographical research literature is that she or he is walking into a conversation which has been going on for the past three decades. Over time, feminist geographers have been developing their own inner debates, vocabulary, theoretical premises, and modes of interaction. When first attempting to wade through this seemingly endless jargon, looking up terms like "feminism" or "poststructuralism" in the dictionary provoke only anxiety. Terms like "paradoxical space" or "positionality" probably won't even be there. We found keeping a small notebook with our ongoing thoughts about what these terms mean useful in developing an understanding of feminist geographical research and providing a path for critique.

7

People Like Us: Negotiating Sameness and Difference in the Research Process

Gill Valentine

Thinking About Feminist Research

Over the last two decades, positivist understandings of objective, impartial, value-free knowledge, in which the researcher plays the role of omnipotent expert extracting information from the passive subject, have been subject to extensive feminist and postcolonial critiques.

In particular, "feminism has challenged traditional epistemologies of what are considered valid forms of knowledge" (WGSG, 1997, p. 87). For example, in the 1980s writers such as Dale Spender (1981) and Liz Stanley and Sue Wise (1983) questioned the notion of universal knowledge, highlighting the way that research done on men had been used to represent all human experiences and that as a consequence women's experiences had been missed out at best and misrepresented at worst. In such critiques feminists also drew attention to the hierarchical power relationships implicit in research relationships, pointing out the ways in which the researcher is positioned as the powerful expert and the respondent as the passive subject (Oakley, 1981).

As a product of these feminist critiques of research feminist epistemologies have emerged (Cope, chapter 3, this volume). These advance alternative forms of knowledge (such as personal experience), highlight the non-neutrality of the researcher, and promote a sensitivity to the power relations which are inherent in the research process (Stanley and Wise, 1983; WGSG, 1997). In particular, feminists have drawn attention to the fact that all knowledge is produced in specific contexts or circumstances and that these situated knowledges are marked by their origins (Haraway, 1991; Harding, 1991). This acknowledgement of the situatedness of knowledge has led to a recognition of the importance of the "position" or "positionality" of the researcher: that we see the world from specific embodied locations (Rose, 1997). This in turn emphasizes the need for each of us to be self-reflexive – in other words to make our own "positions" and the implications of these explicit and known in order to overcome false notions of neutrality (see for example McDowell, 1992a; Mattingly and Falconer Al-Hindi, 1995; Moss, 1995a). England (1994, p. 82) refines this definition of reflexivity as "self-critical sympathetic introspection and the self-conscious analytical scrutiny of the self as the researcher."

This chapter focuses on these notions of positionality and reflexivity. It begins by examining the way that feminists initially employed these concepts, and then goes on to consider some contemporary critiques of them. I then use some of my own experiences as a researcher to illustrate and explore these critiques further. The conclusion reflects on the importance of these debates.

Landscapes of Power?

The notions of positionality and reflexivity have raised many troubling questions about the politics and ethics of whom is entitled to research particular topics. In particular, because of the power relations inherent in the research process some writers have argued that researchers should not work with groups who are positioned as less powerful than themselves, and that to do so is unethical. For example, black women and lesbians have critiqued the way white and/or heterosexual researchers have conducted research on these groups – appropriating their voices and representing their lives (England, 1994), safe in the knowledge that having got "a bit of the other," they can always return to their privileged position as white or heterosexual (Bondi, 1990b, p. 163).

Some of these debates have suggested that researchers who share the same identities with their informants, as for example a woman carrying out research with women, are positioned as "insiders" and as such have *truer* access to knowledge and a closer, more direct connection with their informants than those who are "outsiders." Underlying this notion of being an "insider" is the assumption that they can somehow produce a more correct interpretation of informants' responses. In contrast, "outsiders" are assumed to have more cultural baggage to deal with. For example, commenting on her experience of interviewing women, Janet Finch argued that her interviewees identified with her as another woman. She explains they expected "me to understand what they mean[t] simply because I am another woman" (Finch, 1984, p. 76). She went on to claim that to her it was: "[my] status and demeanour as a woman rather than anything to do with the research process, upon which they based their trust in me" (Finch,

1984, p. 81). Likewise, Marjorie Devault (1990, p. 102) argued that as a woman interviewing women she is able to read between the lines and translate what is being said by the woman she is interviewing. She writes: "I can listen 'as a woman,' filling in from experience to help me understand the things that are incompletely said." She makes these claims on the basis that "women interviewing women is what women have done for generations – understanding one another."

Such claims have been supported by some studies of interview transcripts. Williams and Heikes (1993) studied interviews with nurses conducted by both male and female interviewers, concluding that the informants responded differently according to the gender of their interviewer. This they argued was because the interviewee used the interviewers' gender to gauge the interviewers' opinions and responded accordingly.

However, such understandings of the research process are also problematic because of the way that they (re)produce an insider/outsider dualism and assume that situated knowledge can be accessed through reflection alone. Some of the assumptions outlined above, for example that women interviewing women automatically have insider status, have been criticized as a naïve and dangerous form of essentialism because they uncritically attribute particular connections or a rapport to physical attributes such as sex, or race (Kobayashi, 1994). These dualistic categorizations – insider/ outsider – which are often based on other dualisms such as "white/black," "male/female," "heterosexual/gay" have also been challenged because of the way they obscure the diversity of experiences and viewpoints between and within various groups.

Reflecting on her experiences of interviewing women as part of her work as a research assistant while an undergraduate student, Melissa Gilbert (1994) observes that the commonalities she shared with her interviewees as a woman were undercut by other differences. They were all mothers, Gilbert was not. They were all poorer and had received less education than Gilbert. Some belonged to different racialized groups. As such Gilbert felt very uncomfortable about her material advantages and perhaps rather naïvely tried to negate her privileges and foster a sense of sameness with the women by chain-smoking and making sure that they knew that she too lived on the rough side of town.

Likewise, as part of research with black and white women who have returned to full-time education, Ros Edwards (1990, p. 286), a white academic found that: "I had to take very direct initiatives to place myself as a woman; the black women were not willing to do the placing for me in any other way than race. My placing was not just as a woman but as a *white* woman." Unlike Gilbert, Edwards responded not by trying to emphasize the similarity of her experiences but by locating and acknowledging her difference. By making clear that she was aware that she lacked a shared understanding of black women's experiences of the public world, she found that the women were more willing to talk freely.

Others have had less success at negotiating difference. Kim England (1994) for example has written about her failed attempts as a heterosexual women to complete a research project about lesbian lives. She writes: "Of course, all the sympathy in the world is not going to enable me to truly understand what it is like for another woman to live her life as a lesbian" (England, 1994, p. 86). England's paper describes how her research project encountered difficulties because as a heterosexual woman she was in a more privileged and powerful position than the lesbians she was attempting to work with. These sorts of examples raise troubling issues about questions of power.

In a review of feminist writing about positionality Gillian Rose (1997, p. 313) critiques what she terms the "reflexive landscape of power." She argues that the way feminist writers have conceptualized power relations in the research relationship means that "the relationship between researcher and researched can only be mapped in one of two ways: either as a relationship of difference, articulated through an objectifying distance; or as a relationship of sameness, understood as the researcher and researched being in the same position." She goes on to point out that there is no sense that researcher and interviewee may understand across "difference" or fail to connect through their "sameness." Indeed, notions of "insiderness/outsiderness" suggest that the relationship between researcher and inform-ant can be reduced to social categories such as gender or class and as such these identities are somehow fixed or frozen. Rather writers such as Ann Phoenix (1994) and Bev Mullings (1999) point out that the boundary between the researcher and informant is a highly dynamic and unstable boundary, and that connections and dislocations are about wider biograph-ical moments, much more than social identities such as gender. As such Bev Mullings (1999, p. 4) suggests that researchers need to seek what she terms "positional spaces," "where the situated knowledge of both parties in the interview encounter, engender a level of trust and co-operation." She goes onto suggest that it's better to find such transitory shared spaces that are not informed by identity-based differences because, identities such as gender and class are "rarely fail-safe indicators of an individual's positionality."

Thinking about her own positioning Pamela Moss (1999, p. 158) drew up a list of adjectives: "white, female, high income, ill, vegetarian-who-wants-to-be-a-vegan, and a wannabe blonde." But she reflects they "made no sense without the substance of everyday living that would pull all these positions together into a life, my life. I cannot tell you everything about me." As such, it is impossible for us to unpick how the mutual constitution of gender, class, race, sexuality works in order to identify this as the gender effect or that as the whiteness effect. Rather, as Julie-Kathy Gibson-Graham (1994, p. 219) explains, "I am a unique ensemble of contradictory and shifting subjectivities." Dualisms such as insider/outsider can never therefore capture the complex and multi-faceted identities and experiences of researchers.

Our positionings in relation to our interviewees are never a priori, readily apparent or defined. Rather, they unfold in the course of the encounter (Nast, 1994). The interview is always a joint production. Connections are not made by researcher alone, as researchers we are also positioned by those we work with (Gibson-Graham, 1994). As such we need to recognize that sameness and differences are produced in, and through the research relationship. Drawing on the work of feminist theorist Judith Butler, Rose (1997) emphasizes the relational character of identity – which is always produced through mutually constitutive social relations, arguing that interviewer and interviewee fashion a particular performance of self in interaction. She writes: "social identity is [also] made and remade through the research process . . . What we research is our relation with the researched" (Rose, 1997, p. 315). She continues "She [the researcher] is situated, not by what she knows, but by what she uncertainly performs" (Rose, 1997, p. 316).

In this sense Rose argues that the research process is complex, uncertain and incomplete. Complex because of the diverse intersections of identities and biographies; uncertain because performances might be misread; and incomplete because identities can only be sustained through repetition.

Putting Theory into Practice

For the remainder of this chapter I want to flesh out some of the issues I have outlined so far by drawing on my own experiences – as an "out" lesbian interviewing other lesbians as part of a research project on sexuality; and as a researcher interviewing parents about different aspects of their children's lives where my sexuality was not disclosed. I also draw on the experiences of other researchers to illustrate how identifications and disidentifications are produced in specific performative moments.

As I outlined above, some of the discussion about the research process has talked in terms of fixed identities which are conceptualized in terms of clear and socially recognizable differences such as gender. In practice, the way researchers and interviewees perform our identities and read those of others are uncertain. The identifications and disidentifications we make are complex, with many different notions of sameness and differences operating at the same time. Both researchers and interviewees directly or indirectly claim points of sameness or difference during interviews based not only on knowledge which is exchanged during these conversations but also on what is read off from each others' performances. In other words, as the interview develops, we are constantly (re)producing "ourselves" so that both researcher and interviewee may be multiply positioned during the course of an interview. Ward and Jones (1999) describe the "interview process as a game of positionalities."

For example, during the course of my research on sexuality I interviewed one woman whom I have given the pseudonym Jane. Jane began the interview assuming and establishing a commonality with me as another lesbian. Through our initial conversations about the nature of lesbian relationships and social networks Jane contrasted "us" with her heterosexual friends and colleagues, even though the imaginings into which she embraced me were not necessarily experiences or views which I actually shared. However, as the interview progressed Jane also began to distinguish herself from me, arguing that as a younger and middle-class lesbian able to earn a living from researching sexuality I "had it much easier" than older lesbians like herself. Jane clearly positioned herself as part of a generation who had been at the forefront of campaigns to win recognition for lesbian and gay rights, arguing that "you young dykes don't understand what being gay was like then." Later, the discussion with Jane moved onto experiences of discrimination and violence. When it became apparent that we had shared some similar encounters Jane remembered that earlier she had assumed a generational difference from me, and from then on it became clear that she was uncertain about how to place me.

When I have conducted interviews with parents for research projects about aspects of children's lives the discussions have often begun not from a premise of our sameness – as the interviews with other lesbians did – but from a premise of difference. A difference presumed because I do not have children and in many cases because the respondents also assumed (sometimes mistakenly) that I was younger than themselves.

Like the presumptions of sameness I encountered in the interviews with other lesbians, so too these presumptions of difference were re-negotiated as the interviews unfolded. For example, during the course of my interview with Janet and Brian, a white middle-class couple with two children, the initial presumption of distance between my identity and experiences and their own, which was evident in phrases such as "well, as parents we . . . " soon gave way and it became apparent that I was being embraced into an imagining of sameness epitomized by phrases such as "people like us" and "we." This was not only a product of the fact that we all shared a class and racial identity but also because I fitted in with them in many other ways too. Both Brian and I shared the connection of working in higher education institutions while Janet and I shared some of the same values because of our similar family backgrounds.

However, Janet and Brian's presumption of my sameness was also partially predicated on their misreading of my identity. While I did not deliberately "pass" as heterosexual during the interview, the assumption of heterosexuality in most everyday contexts (Valentine, 1993) is so strong that all the interviewees took it for granted that I too was heterosexual and through my performance I did nothing to disrupt this reading of my identity. In this way, many layers of sameness and difference can be operating at the same time, with the participants and researcher simultaneously identifying and disidentifying with each other, while not even perceiving or recognizing the (dis)connections being made by the other. In Wallman's terms (1978, p. 212) we would each draw "the line of difference" between us in a different place. In turn the ways in which both participants and researchers read each others' identities influences what each feels it is safe to disclose. As a consequence of these complex ways that positionings are (re)negotiated throughout interviews it is impossible to fully know or understand how these shape the knowledge produced from the encounter.

Reflecting on her experiences as a self-identified lesbian and feminist doing a research project interviewing women from the moral right with anti-feminist views, Dona Luff (1999) comments not on the way her identity was read by her informants but rather on her own readings of her interviewees' identities and how these too were fluid and subject to (re)negotiation. Luff (1999) observes that while much is written about how a researcher should dress, and the impact this can have on the way their identity will be read by informants and the consequent interview process, we rarely reflect on our readings of others' bodily performances. Yet Luff found herself reading her informants' identities from their dress. In this excerpt from her diary, she reflects on the confusions that can occur when identities read in this way appear to run counter to researchers' expectations.

> She confused my stereotypes! I unconsciously found myself thinking that she appeared more like a "feminist" than the "moral right" (now what does that tell me? That somewhere I think feminists wear trousers and lots of earrings not printed "frocks" – do I even fit my own stereotype?!) Her home was also very different from the others, but many of her views were actually more conservative than many of the women [who wear nice dresses and have immaculate homes]. I left feeling confounded. (Luff, 1999, p. 694)

Luff goes on to reflect further on the troubling issue of performances, observing that many of the women she interviewed – whom she would in everyday life have dismissed as religious fanatics – were nice people with whom she shared intelligent, thought-provoking conversations. Such that she simultaneously liked and disliked the same person. Indeed, not only did she find her own views challenged, but sometimes she found herself in broad agreement with them.

In a similar way, when interviewing parents about their children for various research projects I too have found myself sharing a sense of connection and warmth with some interviewees, despite the fact they have articulated unpleasant homophobic views (while it is difficult to be positioned by interviewers in ways which are objectionable, Phoenix (1994, p. 57) argues that the articulation of such viewpoints is the whole point of interviews since they are intended to "evoke respondents' " accounts rather than to hear one's own discourses reflected back). In one case in particular, I established a good rapport with, and warmed to, a homophobic couple who were extremely personable and welcoming and we spent a long time after the interview chatting and having a drink. In contrast, I have struggled to develop a rapport and make a connection with lesbians with whom I superficially appear to have much in common. These sorts of interactions show how unstable and shifting interview encounters are and how meaningless static identities and labels are. As Luff (1999) points out people are far more complex than labels such as "moral right" or "homophobic," no interview situation is entirely negative, there is always some moment of connection or rapport.

Indeed, it is worth remembering that a sense of connectedness or sameness does not always prompt the disclosure of thoughts and feelings between the researcher and the interviewee. Rather, it can serve to close down the expression of diverse views as both participants seek to (re)produce the illusion of sameness. In contrast, these openly acknowledged differences in which the interviewee is forthright about their views or challenges the researcher can facilitate more open conversations.

These few examples – of how interviewees have read interviewers and then how as interviewers we might read the performances of our interviewees – question the notion of self-disclosure in the research process. In research encounters the interviewer and interviewee are not locked into static positions described by the usual co-ordinates of class, race, gender, etc. Rather, the way we are positioned in relation to each other is a shifting product of our own fluid performances of the self and the ways that these are read by each other. Although, as interviewers and informants we might think that we choose what we disclose, our performances can always be read against our intentions. As such the way we understand and are understood by each other is always elusive and uncontrollable, and the interview situation unstable and full of ambivalent feelings.

The examples I have outlined so far are fairly conventional examples of conversations and appearances. Ester Newton (1993) points out however that a lot of discussion about research relationships focuses on what is said rather than all the other things that might be going on between the researcher and informant. She then goes on to reflect on the erotics of fieldwork, that when researchers write about ethnography in the first person they talk about thoughts and sometimes feelings, but never desire or lust. The assumption is that sex and emotion between researchers and their informants is absent (the only exception to this being the writing of some women anthropologists about unwanted sexual attention of men or not having sex with men for fear of losing credibility, personal danger, failure of field work, etc.). Newton then goes on to present an account of the relationship she, as a lesbian anthropologist, developed with her best informant. She describes how while carrying out an ethnohistory of a lesbian and gay community in Cherry Grove, a resort on Fire Island near New York City, she fell in love with a woman called Kay. Reflecting on their first meeting she writes, "Was it because I liked her cottage which still had the diminutive charm of an earlier Cherry Grove, because I found her beautiful and her suffering poignant, or because her allusions to past vices intrigued me? Or was it because she called me 'dear' that I came away enchanted?" (Newton, 1993, p. 12).

She then describes how they saw each other daily (over a course of two years) when Newton was in the field, and how their meetings were flirtatious and full of erotic by-play. Kay helped Newton to negotiate access in the field and persuaded people who had initially refused to be interviewed to take part. Newton uses this example to make a broader point about how these encounters shape the research process: "my informants and sponsors have usually been more to me than an expedient way of getting information and something different from 'just friends.' Information has always flowed to me in a medium of emotion – ranging from passionate (although unconsummated) erotic attachment to profound affection to lively interest – that empowers me in my projects and, when it is reciprocated, helps motivate informants to put up with my questions and intrusions" (Newton, 1993, p. 11).

Newton goes on to call for more acknowledgement about our desires for and feelings of repulsion towards our subjects. While in Newton's case the erotics of fieldwork were evident to both researcher and informant, this is not always the case. When I asked a lesbian called Kate whom I had interviewed, and who had also been interviewed by another lesbian researcher called Wendy for a different project, to reflect on her experiences of participating in research, Kate recalled that she had been strongly attracted to Wendy. As a consequence the way that she had presented herself in the interview had been shaped by her self-consciousness about this attraction, and her desire to impress Wendy. Likewise, her reading of Wendy had been undercut by a search for sexual meanings.

It is not only sexual emotions which go unacknowledged in research encounters. Rather, many interviews often evoke some degree of intimacy – because they involve talking about things that are not normally talked about. This intimacy can hold different meanings for the researcher and the researched. I have left interviews that appeared "to go well" feeling distressed; and likewise found some time later that a woman I interviewed was very distressed after the event despite what I understood as our connectedness and the apparent light nature of our conversation because of issues it raised for her which she did not disclose in the interview. As El-Or (1992, p. 65) suggests, there is a "delicate skin overlaying intimacy, separating the overt and the covert, the normal and the abnormal."

Indeed, while we, as researchers have devoted considerable time to attempting the impossible task of reflecting on our own role in the research process we know little about how our informants experience, feel about, or reflect upon their own participation (Ribbens, 1989). Yet, as these examples have demonstrated our informants' experiences and feelings play just as significant a part in shaping the research encounter and the information that emerges from it, as our own. As such there is perhaps a need for some research with interviewees about their experiences of the research process and its outcomes.

The Unknowable and the Not Yet Known

The theoretical arguments about positionality and reflexivity, and the empirical material which I have drawn on in this chapter, demonstrate the multiple positionings and (dis)identifications which are produced and reproduced during the course of an interview. The incoherence of the self – which Rose (1997, p. 314) describes as a "decentred site of differences" – means, as she herself further argues, that the self can never be revealed through the process of self-reflection.

Added to this, the extraordinarily complex nature and general messiness of both the performance, and reading of identities which occur between interviewers and interviewees (as exemplified above), mean that despite all the talk in feminist methodological debates about the importance of acknowledging our positionality or redistributing power between researcher and informant, in reality the research process is beyond the understanding of the researcher. We cannot ever really know what is going on in any given research encounter and therefore how the knowledge we take from it is being produced, nor how the information we use might have been different if our performances, those of our interviewees, or interactions between us, had been different.

This is not to suggest, however, that as researchers we should forget the notion of being reflexive altogether. But rather than attempting the impossible quest of trying to identify a transparent knowable self, our focus should instead be on looking at the tensions, conflicts and unexpected occurrences which emerge in the research process (as exemplified above). By exploring these moments we might begin to decenter our research assumptions, and question the certainties that slip into the way we produce knowledge.

ACKNOWLEDGEMENTS

I am grateful to Pamela Moss for her very constructive comments on an earlier draft of this chapter. I also thank Jenny Kerber for her close reading.

RESEARCH TIP

Personal Presentation in Research Settings

- Wear comfortable clothing.
- Dress in a culturally sensitive manner.
- Dress specific to research context.
- Wear slip-on shoes.
- Don't wear strong scents.
- Don't smoke unless invited to.
- Don't chew gum.
- Wear a watch.
- Turn off pager or telephone, if applicable.
- Check teeth for spinach, poppy seeds, etc.!

STUDY MATERIAL FOR THINKING
ABOUT FEMINIST RESEARCH

WORDS

Agency
Andro-centrism
Authenticity
Cartesian dualism
Cultural appropriation
Deconstruct/deconstruction
Difference
Essentialism/anti-essentialism
Hegemony
Identity
Identity politics
Misogyny
Other
Paradoxical space
Performativity
Political economy
Politics of knowledge
Positioning/positionality
Postmodernism
Poststructuralism
Racialization
Reflexivity
Relativism
Representation
Self
Subjectivity

QUESTIONS

Contemporary theory is often criticized for using too much jargon. The
chapters in this book contain several examples of specialized terms such as
"paradoxical space" and "autoethnography." What are the advantages
and disadvantages of using specialized terminology?

Think about the ways in which you perform your own identity. Are there differences depending upon where you are and who you are with? How does one decide whether a performance is authentic?

Several authors talk about assumptions being made about them as, for example, women or men, as researchers, as sexual beings, and as members of the same social economic class. To what extent do you challenge, play along with, or accept underlying assumptions about difference? About knowledge? About research?

Several of the chapters so far have referred to the high personal and professional costs some feminists pay to do feminist research. For example, there may be less funding than for more traditional projects; there seems to be more pressure to explain feminist research and justify projects; and there may need to be more time for the research that produces fewer traditional products, for example, publications. Would you conduct feminist research in the climate where you study? How public would you be about your work? What issues do you need to consider while making such decisions? What is at stake for feminists who undertake or don't undertake feminist research in geography?

ENGAGED EXERCISE

Every act of translation implies some degree of alteration and interpretation of meaning. Cross-cultural research poses specific challenges to the communication of information, not only because of linguistic differences, but also because of the variations in conceptual categories that play out in language. Meanings can be distorted when the conceptualization behind words does not correspond with another person's or culture's intended meaning. Variations in meaning may be determined both spatially and linguistically. For example, a single plant can have many common names, or might have even more than one scientific name. We illustrate this point by recounting a research experience of one of the members of our Working Group, whose research team is concerned with the use of native plants by a First Nations group in the interior of British Columbia. After conferring with elders about their knowledge about and the location of the red-osier dogwood (*Cornus stolonifera*), the research team was perplexed that the elders were unfamiliar with this plant. The red-osier dogwood is commonly found along waterways in the interior of British Columbia, and is known to be of cultural significance to the peoples in this region. After completing the first round of interviews, the team learned that the name for *C. stolonifera* in that area is "red willow." As one of the positive results of this miscommunication, the research team now knows the *appropriate* terminology for phrasing a useful research question particular cto this location.

This is but one example of the difficulties with translations and cross-cultural research. To explore the nuances of translation first hand, you might try the following exercise using the website http://www.babelfish.altavista.com/, a program set up to translate words and phrases from one language to another – often with startling results.

Write a brief paragraph in which you wish to convey a particular meaning. Enter the paragraph into the website as instructed, then translate it first into another language, then into a second, and finally back to the original. How has the meaning of the original paragraph been enhanced, distorted, lost, or altered in the process of translation?

For other sites concerned with language translations, interpretations, and word associations, see also:

http://www.thinkmap.com
http://www.plumbdesign.com/thesaurus/
www.thesaurus.com
www.onelook.com/
http://www.travlang.com/languages/

SUGGESTED FURTHER READING

Thinking about research in feminist geography requires going beyond writings specifically about methods and into writings challenging the ways in which knowledge is produced and power is deployed and resisted. Topics addressed throughout this section cover a wide range of methodological issues as well as feminist conceptualizations of theory, power, and knowledge. Suggestions for reading will necessarily fall short of expectations precisely because the implications for research design, the conduct of research, and analysis are never exhausted, nor are the "answers" very clear. Bear in mind, then, this list is only a series of suggested entry points into extensive literatures outside methodology *per se*.

On gender and space, see:

Bondi, L. 1990a: Feminism, postmodernism and geography: space for women? *Antipode* 22, 156–67.
Laurie, N., Smith, F., Bowlby, S., Foord, J., Monk, S., Radcliffe, S., Rowlands, J., Townsend, J., Young, L., and Gregson, N. 1997: In and out of bounds and resisting boundaries: feminist geographies of space and place. In Women and Geography Study Group, *Feminist Geographies*, London: Longman, 112–45.

Massey, D. 1994: *Space, Place, and Gender*. Minneapolis: University of Minnesota Press.
Rose, G. 1993: *Feminism and Geography: The Limits of Geographical Knowledge*. Minneapolis: University of Minnesota Press.

On the politics of identity, see:

Bondi, L. 1993: Locating identity politics. In M. Keith and S. Pile (eds), *Place and the Politics of Identity*. London and New York: Routledge, 84–101.
Jones, J. P. III, Nast, H. J. and Roberts, S. M. (eds) 1997: 'Part One: Difference', in *Thresholds in Feminist Geography*. Latham, MD: Rowman and Littlefield, 1–115.

On black women's writings, see:

Collins, P. H. 2000: *Black Feminist Thought: Knowledge, Consciousness, and the Politics of Empowerment*. Revised tenth anniversary edn. London and New York: Routledge.
hooks, b. 1994: *Outlaw Culture: Resisting Representations*. London and New York: Routledge.

On racialization, see:

Anthias, F., Yuval-Davis, N. with Cain, H. 1992: *Racialized Boundaries: Race, Nation, Gender, Colour, and Class and the Anti-racist*. London and New York: Routledge.
Anzaldúa, G. 1987: *Borderlands/La Frontera: The New Mestiza*. San Francisco: Spinsters.
Kobayashi, A. and Peake, L. 1994: Unnatural discourse: "race" and gender in geography, *Gender, Place and Culture*, 1, 225–43.

On racialized colonial relations, see:

Alexander, M. J. and Mohanty, C. T. 1997: *Feminist Genealogies, Colonial Legacies, Democratic Futures*. London and New York: Routledge.
Mohanty, C. T., Russo, A. and Torres, L. (eds) 1991: *Third World Women and the Politics of Feminism*. Bloomington: Indiana University Press.

On performativity, see:

Butler, J. 1990: *Gender Trouble: Feminism and the Subversion of Identity.* London and New York: Routledge.
Butler, J. 1993: *Bodies that Matter.* London and New York: Routledge.
Gregson, N. and Rose, G. 2000: Taking Butler elsewhere: performativities, spatialities and subjectivities, *Environment and Planning D: Society and Space*, 18, 433–552.
Nelson, L. 1999: Bodies (and spaces) do matter: the limits of performativity, *Gender, Place and Culture*, 6, 331–53.

On masculinism, see:

Berg, L. D. 1998: Reading (post)colonial history: masculinity, "race", and rebellious natives in the Waikato, New Zealand, 1863, *Historical Geography*, 26, 101–27.
Jackson, P. 1991: The cultural politics of masculinity: towards a social geography. *Transactions of the Institute of British Geographers*, 16, 199–213.

On embodiment, see:

Butler, R. and Parr, H. (eds) 1999: *Mind and Body Space: Geographies of Illness, Impairment and Disability.* London and New York: Routledge.
Moss, P. and Dyck, I. 1996: Inquiry into environment and body: women, work and chronic illness, *Environment and Planning D: Society and Space*, 14, 737–53.
Nast, H. J. and Pile, S. (eds) 1998: *Places through the Body.* London and New York: Routledge.
Teather, E. K. (ed.) 1999: *Embodied Geographies: Spaces, Bodies and Rites of Passage.* London and New York: Routledge.

On hegemony, see:

Laclau, E. and Mouffe, C. 1985: *Hegemony and Socialist Strategy: Towards a Radical Democratic Politics.* London: Verso.
Mouffe, C. 1992: Feminism, citizenship and radical democratic politics, in J. Butler and J. W. Scott (eds), *Feminists Theorize the Political.* London and New York: Routledge, 369–84.

On the articulation of poststructuralism and feminism, see:

Nicholson, L. J. (ed.) 1990: *Feminism/Postmodernism.* London and New York: Routledge.

Pratt, G. 1993: Reflections on poststructuralism and feminist empirics, theory and practice, *Antipode,* 25, 51–63.

On anti-essentialism and analysis, see:

Gibson-Graham, J.-K. 1994: "Stuffed if I know": reflections on post-modern feminist social research, *Gender, Place and Culture,* 1, 205–24.

Spelman, E. V. 1988: *Inessential Woman: Problems of Exclusion in Feminist Thought.* Boston: Beacon Press.

On institutional power and knowledge, see:

Smith, D. 1990b: *The Conceptual Practices of Power: A Feminist Sociology of Knowledge.* Boston: Northeastern University Press.

Mariniello, S. and Bové, P. 1998: Gendered Agents: Women and Institutional Knowledge. Durham, N. C.: Duke University Press.

Part III

Doing Feminist Research

Decentering Authority!

Feminist Pegagogy Working Group

For many of us, our experiences in learning have led us to believe that the content of the books we read are true, that the *authors* are knowledgeable, and that somehow together they constitute *authority*. During our school days we were required to read textbooks and then tested on our knowledge of "facts." If we remembered what was written by the author(ity), we were rewarded with high marks; if we responded with something different, something other that what the author(ity) decreed, we received low marks. If our answers were wrong, then, as the authoritative logic goes, we were wrong too!

Being "wrong" has its advantages. Like with any hegemony, the authority of doing research included, we give way to dominant ideas that are written by authors with authority and both knowingly and unknowingly assist in perpetuating such ideas. We get caught up in routine complacency, failing to question authors and authority. We end up repeating things we have learned through reading some book, citing some article, or repeating some argument and implicitly reproducing the same configurations of authority within a particular field (feminist methodology in geography included). Yet at the same time as we are entranced and seduced by this dominance, we know that there is space to resist power, to challenge hegemonic stability, to contest authoritative writing. We don't have to believe that the only good research is research that is value-free, scientific, and reproducible. We don't have to validate our standing as researchers by conforming to the academic standards based on a set of values feminists have shown not to be useful for research topics associated with power relations that oppress and marginalize individuals and groups of people. And we don't have to use the same research methods that have been deemed valid through generations of authors writing about research claiming authority in the field of doing research.

In fact, we really have not forgotten those wise and well-intentioned, albeit imperious and authoritative, words: "Don't believe *everything* you read!" But moving toward a corollary, "Don't believe *anything* you read!" may be too harsh and cynical because what really is at issue is not the authority itself, but the process of creating authority. Authority involves questioning who speaks, who gets to be heard, and who decides each. Assessing authority involves focusing on knowers, forms of knowledge, and what is and can be known as well as on identifying and evaluating truth claims – those that write and represent the world. And, perhaps most significant, is the context within which authority plays itself out – the power that each of us brings to bear as multiply positioned "subjects" within interwoven relations of power.

Our ideas about authority, as laid out here, infused our discussions about the material in this book. What could we include? Could we have an on-line learning environment with all the authors there for interaction and debate? Could we create virtual classrooms with group projects with feminists across the globe? What should we include? Should we list the articles that we found most useful in getting us to think about feminist methodology? Should we share our strategies about how we learned to do feminist research? Although we wanted to break that pattern of conventional texts that laid out word lists and questions for definition and discussion, we realized that this was indeed the way we learned. We got to know what issues we wanted to address and what tasks we wanted to undertake through working through exercises with each other in and out of class.

As practicing feminists, all of us have shuddered to think that because we have *taken* this opportunity to participate in contributing to a pedagogical project about feminist methodology in geography, we might be considered authorities – as a collective who knows because we've been there, we've done that, and we've figured it all out. At times, we nearly succumbed to the (negative) notion that we would be taken as an authority and discussed what we thought the best way for students to learn how to undertake feminist research in geography. At the same time we knew that despite our "best intentions" we were complicit in silences and erasures because our assumptions and arguments could not be removed from the same power-laden contexts within which dominant accounts of knowledge – like this textbook – are produced. We struggled with our angst over whether we would be considered hypocritical and arrogant because we dared to participate in developing a feminist textbook on research. Most of us, at some time or another, anguished over these issues enough to think about withdrawing from the project.

But, as you already know, we did seize this opportunity to write our bit of the world as we saw it. We realized that we faced many of the same dilemmas that the feminists writing for this text did. Our quandaries arose from our experiences in classrooms and in our everyday lives and we decided that we wanted to be among those involved in working toward decentering authority, toward destabilizing dominant masculinism that continues to permeate teaching and learning, toward feminist transformations of the classroom. So, even though we have selected what to (re)produce in the learning materials, we did so with the realization that readers would approach these words, information, and ideas with caution and scrutinize them with a healthy dose of skepticism.

The authors of the chapters in this section are concerned primarily with particular methods associated with the actual "doing" of feminist research. None however professes to have absolute confidence in or knowledge about the choices they have made while doing feminist research. Each struggles, at least implicitly, with the authority of not only her own work, but also the work of feminist researchers in geography collectively. Karen Nairn recounts the choices she made about putting together a research project about data collection methods and design. She makes the point that accessing similar information can come in many forms just as one piece of information can hold different types of information. Mei-Po Kwan links the use of quantitative methods to feminist research. She shows through hands-on examples that feminist uses of quantitative data and analysis are not only possible, but also necessary for some research questions. Joan Marshall discusses several dilemmas she encountered while engaging in ethnographic research with a closely knit community where she was an outsider. She does not place the choices she made as solutions to problems, but rather as resolutions to situations. Deirdre Mackay offers an intimate look at the process through which she came to understand how efforts at reciprocity were offered and (possibly) received. She prompts us to think about research as a "work in progress." Kim England provides a wonderful account of "studying up" and points out how research is not a smooth process. She argues that research is not about the perfect or complete project, but rather about "the art of making do." Geraldine Pratt writes about focus groups as a way to "decenter" the authority/centrality of the researchers and the practice of research. Although focus groups may offer the potential for less hierarchical relationships, Pratt reminds us that they never disappear. Together these chapters spell out only a fraction of methods available to feminist geographers undertaking research and of the dilemmas they have or will face in the field.

Short 3 Doing Geography as a Feminist

Reconsidering Success and Failure in Feminist Research

Maureen G. Reed

It didn't get anybody a full-time job, it didn't give a forest worker a job, it didn't contribute to the renewal of our forests. ("Brenda," cited in the *Squamish Chief*, a local newspaper, about the outcomes of my research project in Squamish, Canada, 2000)

Thank you Maureen. Your concern is felt. (Inscription on a gift to me from the non-governmental organization, North Island Women, Canada, 1999)

This is an article written from a sense of failure. (Rose, 1997, p. 307)

I face these quotations with grave ambivalence. I am ambivalent because each quotation represents a different audience for research; each audience holds different standards for the evaluation of its "success" and "failure." I undertook a research project that was intended to provide results for public policy related to forestry practices and management of public lands. It was not a topic that readily lent itself to feminist research. It was my first project with an explicit feminist orientation; one of only two that would be funded by this multi-million dollar research program. This combination – public policy focus of the research, my lack of prior experience in feminist research, the presumed dubious relevance of feminist research to the aims of the funding agency, and my own desire to "make a difference" both theoretically and in the lives of my research subjects – created for me a near-impossible challenge for assessing the outcomes of the project. Hence, my ambivalence. Upon reflection, I have become most disturbed by the challenge of meeting the dual needs of academic scholarship with the demands imposed by public policy. I am grave because I believe that this quandary is likely to be reproduced again and again as the funding of social science research becomes more heavily reliant on both public and private sources. As part of being in this quandary I ask myself, "how might research by feminist geographers be affected by these changes?"

Notice how I juxtapose the "needs" of scholarship with "demands" of public policy. I could have easily reversed these descriptions. I could have discussed the challenge of meeting the demands imposed by scholarship with the needs of public policy. Why do I consider the public policy aspect of research a demand imposed from outside my own research agenda? I think there is more to it than simply reasoning that I am situated *within* an academic environment. Rather, my positioning reflects my biases about what I believe are the central aims and limitations of both the academy and policy environments, and I am situated within changes in feminist scholarship itself. If feminist scholarship is rooted in attempts to explain *and* to change unjust social relations, then feminism has a direct application to public policy. Yet, feminist scholarship in geography has increasingly moved towards more theoretically informed, academically approved standards for judging research results (McDowell, 1992a; Bondi, this volume). What does this say about how far we have become entrained within academic conventions for judging our successes and failures? Perhaps more importantly, how can we be directly engaged in these messy, real-world debates without compromising the intellectual integrity of scholarship?

Funding agencies require feminist geography research to be policy-relevant. This demand widens the range of adjudicators and shapes research expectations for *our* projects. These interconnections exist at all phases of the research process, from the formulation of a problem through the generation of research questions, from the selection of methods and entry into the "field" through analysis of results and their articulation and circulation in both academic and policy forums. I view this as something of a Faustian bargain – by accepting funding from such agencies, what compromises might I have to make with respect to my own scholarly integrity and feminist intentions? While feminist researchers have examined their own positionings in relation to their research subjects (e.g. related to similarities or differences of gender, class, sexuality or ethnicity), they have been less attentive to positionings in relation to dynamics of the context of the research – the tensions that arise between different expectations of the academy, funding agencies and how the relationships between the two are played out in setting priorities for research and criteria for evaluating results.

Setting Research Priorities and Practices

New research priorities are beginning at the outset to shape our research questions and methods. For example, the major publicly funded social science funding agency in Canada, the Social Sciences and Humanities Research Council (SSHRC), historically has provided research grants that have not required a public policy focus as a criterion of relevance. But times are changing. For individual researchers, SSHRC grant applications now *require* applicants to identify how results will be "disseminated" and made relevant outside the academy. For larger, collaborative proposals, applicants are required to establish partnerships outside academia and to demonstrate how academics will contribute to the aims of their "lay" or "community" partners. Increasingly, and not surprisingly, Canadian researchers are seeking funding outside SSHRC. Funding may be obtained through a variety of agencies such as government departments or ministries and crown corporations as well as non-governmental organizations such as credit unions or advocacy organizations. All these agencies demand that their own policy priorities be reflected in the proposals if they are to be successfully funded, which becomes problematic quite quickly. These agencies may pre-define key topic areas for research. They may pose direct questions for academics to answer. They may accept some methodologies, reject some, and appropriate still others. My research examined the social lives of women living in forestry-based towns on the west coast of British Columbia (BC). Some of these were on Vancouver Island; one was located in Squamish (near Vancouver). My focus was aimed squarely at those women who continued to support industrial forestry in light of substantial challenges to the industry and communities – challenges from economic restructuring, environmental activism and recognition of the rights of Aboriginal First Nations. I secured funding from Forest Renewal BC (FRBC), a newly created crown corporation in British Columbia whose primary objectives were to improve the well-being of BC's forests, forest industry, and forestry communities. To obtain funding, I had to demonstrate how my study would improve the well-being of forestry communities overall. My initial proposal was accepted, pending my agreement to revise its objectives according to the demands of a single reviewer. This reviewer requested that I expand the group of women beyond "protesters" to include a sample of women from a broader political and economic base. Thus, my compromise was to carry on with my own interests, and "append" those of the reviewer, in order to satisfy the granting agency. But it didn't stop there: I had to do this every time I made a report about the project (I made fourteen (!) in total), hampering my abilities to stay true to my own research interests and questions.

The funding from FRBC also raised the expectations and hackles of residents from forestry communities themselves. The provincial government told residents that monies from this corporation would be used to support local development of forest resources and communities. They were hopeful about research but also wary – concerned that funding for research would supersede their own needs for assistance during times of massive economic restructuring. Although FRBC administered two separate programs – a research program and a land management program (e.g. funding activities such as tree planting and watershed restoration) – in the minds of many community residents, the objectives of the separate programs were blurred by their pressing concerns about the survival of local livelihood, community and culture within a re-forming forestry economy.

Notwithstanding these limits, there were some important benefits of going this route. The research grant from FRBC was about three times larger than what I could have secured through SSHRC. This funding, therefore, was like receiving gold coins from heaven. It enhanced my own status within the department and university. As Demeritt (2000) points out, the amount of grant funding is an increasingly important currency in measuring the value of an academic. My relatively large grant greatly increased my average income to the department and university. While large grants are valued, few individual feminist research projects can claim such advantage. More importantly, the three-year award allowed me to pay generous honoraria for women to take part in each workshop and to conduct interviews. Thus, I provided "fair wages" for women to work on the project, so that they would value their own contributions and know that I did so as well. In addition, I was the first person at the university to officially reimburse non-university women researchers for child-care expenses incurred while they were occupied in the research project. In my mind, these were small, but important successes. They were part of my contribution to redistribute the wealth provided to me through this grant.

Judging the Result: Expanding Assessment Criteria and Critics

A movement into the realm of public policy research also expands the networks of relevant actors who judge the research effort – from the point of application, through its implementation, and ultimately upon its completion. For example, usually only the most obscure research proposals funded by the government get cited in the newspaper each year in public jests about academic irrelevance. In contrast, the explicit public policy orientation of the FRBC project expanded the range of interests, relevant criteria, and potential critics pertinent to assessing its methods and results. This research was set within a policy environment where forestry was a major provincial industry undergoing economic restructuring and government agencies were looking for research with policy relevance to help make governmental decisions. For some community research participants, accountability and research relevance meant the power of the academic to generate positive change in their lives. Yet at the same time, expectations and prospects for failure were undeservedly high. This was brought home to me when I listened to one of the transcripts involving the trained women researchers:

> *Participant*: Is that what's going to happen with this [research . . . that is, to be put on a shelf and ignored]? . . . I'm just curious.
>
> *Interviewer*: I think Maureen has a personal interest, not a vested interest, but a personal interest in the results. I believe, from meeting her, that she has a real interest in this sort of thing and she has an incredible amount of valuable knowledge and I do believe that she's got, in her position as being professor at [The University of British Columbia], she's probably got a high, a tremendous amount of respect too . . . It's been my experience that people who are in educational institutions have credibility and knowledge and knowledge is power, so, you know. Hopefully, and as I say, she does have a personal interest in it, from the heart, not from . . . (*interrupted*).
>
> *Participant*: Perfect, that'll be good.

Here, the interviewer appears to have a much inflated view of my power to change policy. She translates my "concern" into a measure of success. Yet, from the perspective of the most important criterion of this public policy context – jobs – the project was an utter failure: no jobs were created! In another interview from Vancouver Island, "Betty" remarked:

> And [another friend] was saying, well, ask why they're doing it too, because she thought it was all a bit peculiar that this was going on, and for so long, and money going to this [research] when it's, it's been cut for the men working who could maybe use the money . . . I guess, we just have trouble, a lot of trouble with, I guess, government spending and what money will go to and what it doesn't go to. It just doesn't make a lot of sense sometimes.

This quote dovetails with Brenda's assessment from Squamish that I highlighted at the beginning of the paper. Both continue to haunt me. They haunt me not because I intended to generate or supplant job opportunities from the research, but rather, because this public expectation permeated the project despite my assertions to the contrary. This expectation was entirely reasonable because the agency that funded the research had also made such a strong commitment to job creation. The fact that the research program was separate from other programs of FRBC was not relevant to those living in the communities. If I ignored this point, I only highlighted my difference and my distance from my research subjects – that of a detached, urban, privileged, useless, *yet employed*, academic. If I embraced it, I set myself up for failure. My research simply did not create jobs. Perhaps, indeed, Betty was right.

A Feminist Scholarly Context

In keeping with other feminist scholars, I acknowledge that the power to define the parameters and to impose measures of success are not mine alone. Presumably the granting agency can use the reports and findings in ways that I had not intended. I have not monitored the follow-up of all the reporting requirements (meeting the requirements was draining enough). Importantly, for this essay, the research subjects themselves can shape the criteria for success, apply them, and announce them in unexpected ways. Just before final reports were made public, Brenda was elected as a municipal councilor. While I was aware of Brenda's skepticism from the outset of the project, I did not predict her political aspirations. Her election as a local politician gave her a platform and a public legitimacy that she did not have before. While I could argue that her involvement in the research helped to hone some necessary political skills and community insights to improve her own election prospects, these arguments sound like griping. She certainly did not attribute any part of her success to any enhanced training or information she received through the project. I discovered her comments in the local Squamish paper some time after they were made. She had not mentioned her involvement in the research. I was certainly in an unequal position to debate her position. Nor would it have been appropriate to contradict her.

This acceptance of others' power within the research process is consistent with the observations of feminist scholars who point out that researchers are situated subjects who only have partial knowledge and power to inscribe the research agenda. Rose (1997) described the power relations between researched and researcher as fluid marked by fragmented understandings, uncertainty, and risk throughout research and dissemination practices. She suggested that "the research process is dangerous . . . [where] the risks of research are impossible to know" (Rose, 1997, p. 317). I could not have known, nor even anticipated, that Brenda would turn against me publicly.

In contrast to Rose's more negative assessment, Demeritt (2000) argued that as identities of and relations among researchers, partners, and publics are changing, the processes of "trust building, mutual understanding, and social learning involved in doing research can be as important for participants as the substantive results" (Demeritt, 2000, p. 326). I agree that these intangible effects are often discounted. The fact that my concern was felt suggests some measure of success, at least on Vancouver Island. The invitation to be the keynote speaker at an inaugural celebration of "Women of Influence" on northern Vancouver Island suggests that during my research, I attained some level of local trust and support. Yet, I am acutely aware that mutual understanding does not feed a family. And when research does not feed a family in a public policy context where communities believe they are threatened by actual or imminent (nutritional or political) starvation, processes of trust-building erode. So, too, erodes the perceived public value of the research project.

Not a Conclusion

While I have focused on a single project and funding agency here, I use this example because I believe it has wider application. Academics are now encouraged by university administration to go to non-conventional funding sources that may have strings attached to public policy objectives. Even conventional sources such as SSHRC have new requirements for researchers to express their relevance to society. Feminist scholars, who try to uncover and undo real-world inequities, and who have focused scholarly attention to power relations should be at the core of debates about funding. These arrangements shape all aspects of a research project – from its initial conceptualization of the "problem" at hand, through the choice of methods, analytical strategies, and interpretation of results. Importantly, in the latter categories, our successes will be determined in very public and uncontrolled arenas of policy debates and study locations. More fundamentally, these arenas shape the basic premises of the research itself. Sometimes they do so overtly, such as in pre-establishing questions for research programs. Sometimes these arenas shape research more subtly through ongoing renegotiations of the research project through its various stages. As feminist scholars, we need to discuss the implications of this emergent research context fully and openly among ourselves, policy makers, and research subjects.

RESEARCH TIP

Pre-interview Preparation

- Do your homework, e.g. review contact material, review purpose of material, review relevant/appropriate documents.
- Do a practice interview.
- Arrange for interpretation or translation, if necessary.
- Confirm appointment time and place.
- Prepare interview bag.
- Make sure you know where you are going.
- Leave interview destination and details with a friend or a colleague.
- Take time to collect your thoughts just before the appointment.
- Don't arrive in a rush.
- Be prepared for the unexpected!

8

Doing Feminist Fieldwork about Geography Fieldwork

Karen Nairn

I am a geographer who has chosen other geographers – geography students and geography teachers/lecturers – as the subject/object of my research. I investigated the social conditions of the reproduction of geographic knowledge in the context of residential fieldtrips. Fieldtrips are defined as trips away from an educational site and may be one hour or one day long. Residential fieldtrips involve students staying away at least one night. The teaching and learning of fieldwork practice is a significant aspect of the reproduction of geographic knowledge because many geographers claim it distinguishes geography as a discipline (see for example Stoddart, 1986). Although other geographers critique geographers' preoccupation with particular styles of fieldwork (see for example Berg, 1994; Nairn, 1998), the residential fieldtrip is one forum through which the practice of fieldwork is intensively taught.

In my recently completed doctoral dissertation (Nairn, 1998), I demonstrate how particular forms of collective masculinity dominate the social environment of university residential fieldtrips. A critical analysis of *how* fieldtrip cultures are part of generating particular forms of geographic knowledge also provides an opportunity for engaging a feminist methodology that could "be a form of resistance to dominant ways of acquiring and codifying knowledge" (Nagar, 1997, p. 203). In order to undertake a feminist analysis of the reproduction of geographic knowledge in the context of residential fieldtrips, I collected data based on a feminist principle that a different gendered "reality" is possible. I focused "on the changeable, marginal, deviant aspects – anything not integrated which might suggest fermentation, resistance, protest, alternatives – all the 'facts' unfit to fit" (Gebhardt, 1978, p. 405). Madge et al. (1997, p. 90) argue that feminist methodologies are distinguished by their "challenge to research orthodoxy" and identify four characteristics of these methodologies – ways of knowing, ways of asking, ways of interpreting, and ways of writing. In this chapter, I draw on my dissertation research and recount what I found as significant in ways of asking and ways of interpreting.

The Politics of The Field

My stated feminist theoretical and methodological concerns sparked the politics of the field, and in turn the field shaped my theoretical and methodological directions. Theory and practice, knowledge and politics are interconnected because "knowledge and its production [is] *always already* [a] political process" (Gibson-Graham, 1994, p. 214; emphasis in original). Fieldtrips and fieldwork, the "heart" of what (academic) geographers do (Head of Geography Department, fieldnotes, October 19, 1994), were not considered a legitimate doctoral topic at one university in Aotearoa/New Zealand. For example, one academic geographer who responded to the initial doctoral proposal asked "Is it geography? Why geography? Why not education? . . . certainly not the sort of research usually done in geography" (fieldnotes, November 23, 1994; see also Kobayashi and Peake, 1994, p. 239, and McDowell, 1992a for further discussion). My position as student, combined with my positions as woman, feminist, and ex-high school teacher, cast me as someone without the authority to ask questions about and carry out research on the teaching practices of academics. McDowell (1992b, p. 59) points out the dilemma I was facing – "it is difficult to simultaneously be seeking validation from and critiquing the academy."

From a list of possible questions, I distilled one key research question: How are the embodied disciplinary identities of geography student/geographer socially constructed and reproduced through the culture of residential fieldtrips in Aotearoa/New Zealand in the 1990s? In other words, how do geography students learn to be geographers and to do geography? The term "embodied disciplinary identities" specified both the bodily and the disciplinary as two key dimensions of residential geography fieldtrips for scrutiny. Such concerns intersect with feminist interests with embodiment and identity (see for example Davis, 1997; Grosz and Probyn, 1995; Longhurst, 1995). At the time I formulated the topic there was no sustained critique of geography residential fieldtrip education in Aotearoa/New Zealand.

The study included seven residential geography fieldtrips (ranging from two to seven days) of two high schools and two universities located in Aotearoa/New Zealand (Nairn, 1998). I focus more attention on the three residential fieldtrips of two universities because the university is a key site of knowledge (re)production via teaching and research. The culture of universities is one where academics are usually the researchers, not the researched, and the disciplinary culture of geography is one where geographers are used to observing other places and people, not themselves and their education practices.

It was also important that the research move beyond critical inquiry to a reconsideration of what could be different if geography fieldtrips were not so central to the geography discipline and/or were conducted in alternative ways. I argue for a new metaphor for fieldwork. What would fieldwork knowledge be like if *reconsideration*, rather than discovery/exploration which has long been the metaphor of enlightenment (Myerson and Rydin, 1996), was adopted as an underlying principle for fieldwork? The metaphor of discovery is caught up in the privileging of "seeing the real world." The metaphor of reconsideration enables the basis on which knowledge is claimed, and the structures of that knowledge, to be interrogated.

Ways of Asking, or Data Collection

"Every method has its shortcomings" (Gilbert, 1994, p. 95), so I used multiple qualitative methods of data collection to compensate for some of the shortcomings of each of the respective methods (Denzin, 1978; D. Rose, 1993). The use of multiple methods also increased the likelihood of uncovering any "facts" unfit to fit. I chose each method of data collection for its appropriateness to the respective phases of the research process, in particular to the level of empathy that could realistically be expected at key points, as well as to the material conditions (classroom, lecture theatre or fieldtrip location) of each stage of the research. For example, the beginnings of trust and empathy from the participant observation phase (where I lived and worked with participants) facilitated the interview phase, when I expected to collect more in-depth data from selected participants.

There were four different data collection phases. I designed a pre-fieldtrip exercise, which included a drawing task, to record students' expectations of the upcoming fieldtrip. I used participant observation during the fieldtrips. At the end of each fieldtrip I asked students to complete a written evaluation of the fieldtrip. Some months after the end of the fieldtrip, I conducted in-depth interviews with selected fieldtrip participants. Thus ways of asking included drawing, writing, listening, observing and participating in fieldtrips. In the process of arguing for a more diverse repertoire of ways for geographic knowing, I wanted to mirror such theoretical contentions in the methodology. Or, put more prosaically, I wanted to practice what I preached.

As part of engaging in reconsideration, in this chapter I pay attention to how each of the methods offered the potential for alternative ways of asking within a disciplinary paradigm patterned by ways of asking that privilege text, seeing for yourself, and forms of disembodiment. And threaded through the discussion is critical reflexivity (or reconsideration) that is a hallmark of feminist research (see Bernstein, 1992; Lather, 1991; McDowell, 1992a; Rose, 1997). Reflexivity enables assessment of research theory and practice, of what does and does not work so that adjustments can be made to ways of asking (and/or ways of interpreting). Rose (1997) cautions against reflexivity as a new feminist orthodoxy but in the process of being reflexive about reflexivity, she implicitly demonstrates the importance of ongoing vigilance about the politics of research.

I chose both drawing and writing media to provide students with more than one way in which to express themselves. I was interested in "key words, concepts and images, as well as apparently disjunctive free associations" (Okely, 1996, p. 40) and the drawing method fulfilled my goal of facilitating spontaneous associations with the idea of fieldtrips. The drawings also signalled important theoretical issues. There were drawings of body parts, whole bodies and predominantly stick bodies that sent me off on important theoretical tangents. This way of asking provided one alternative to the privileging of written text but engendered challenges for ways of interpreting that I describe in the next section on data analysis.

Participant observation is a method that "delve[s] beneath the surfaces of observed phenomena in order to seek the meanings and intentions which produce it" (Evans, 1988, p. 199). By participating in social phenomena we observe, we are more likely to learn the underlying meanings which produce that phenomena (Evans, 1988). I wanted to do more than observe for three reasons. First, I was already critical of geographers' over-reliance on observation and appearances as a measure of reality and/or truth (Cloke et al., 1991; Rose, 1992). Second, the appearance of a social phenomenon does not in and of itself provide an explanation of that phenomenon. Participation offers another point of access to the meaning of that particular phenomenon (although it does not guarantee access and/or understanding). Third, "the fact of participation, of being part of a collective contract" (Evans, 1988, p. 209) is an appropriate method for investigating the nature of the collective contract. A fieldtrip is a relatively unique collective contract in which participants live with people they have previously sat alongside in a lecture theatre: "they're people that you don't know, I normally wouldn't choose them to go away on a holiday with, or whatever" (interview with a male student after a university fieldtrip, September 15, 1995). Participation enabled me to "come across" the terms of belonging to a collective of geography students. The terms of belonging included many elements, for example, dress (shorts or jeans and polar fleece jackets), drink (beer rather

than wine), physical ability (walking long distances) and social ability
(staying up late in spite of early starts the next day).

Participant observation enabled me to access embodied forms of knowing more readily than any other form of data collection. Participation
alongside students in the material conditions of each fieldtrip provided
access to embodied knowledge such as the taste of fieldtrip food, the
temperature and comfort of sleeping accommodation, the physical demands
of long days and the social demands of relating to large numbers of fieldtrip
participants. This meant that I collected data about my own as well as
about other participants' embodied experiences. These data I often gleaned
from comments I overheard at mealtimes, during fieldwork activities, and
in the bunkrooms that I shared with students and staff. It was important
to include myself as subject of data rather than collect data solely about
the "object" of research because I was conscious of feminist critiques of
disembodied researchers who are absent from their research. Although
participant observation offered the most potential for embodied forms of
knowing via a greater sensory repertoire and for countering the privileging
of the visual via the participation component, it also felt the most intrusive.

As a researcher who lived and worked with the subjects of my research
I had access to the private dimensions of (some) fieldtrip participants' lives.
Could I (or should I) stop being a researcher in some contexts? Where
would I or should I draw the boundary between the information that they
consciously told me as a researcher and the information that they might
(inadvertently) provide during a casual conversation in a social/informal
context? At times, it was hard to go to bed during fieldtrips, knowing that
I could miss out on research opportunities. I was interested in the socializ-
ing aspects of geography residential fieldtrips, yet it seemed at odds to be
working (researching) during the social activities of fieldtrips. I developed
the art of making a can of beer last a long time so that I would have an
appropriate prop in a social situation on a university fieldtrip, and continued my work. I felt like I was exploiting social situations for work
purposes.

This sense of exploitation was complex. I was representing research
subjects who held more power as well as those who "held" similar or less
power than myself. I aim therefore to "not make public information or
strategies that may compromise the less powerful" (McDowell, 1992b,
p. 408; see also England, 1994; Katz, 1994) but I do not apply this same
criteria to those in positions of power. This point requires further qualifi-
cation. I do not name those in power but information about the practices
and strategies of those who exercise disciplinary power are included in
written products because the purpose of my research was to explicate the
disciplinary culture of geography fieldtrips, a culture shaped by those with
the power to do so.

I nevertheless sought ways to minimize the effects of my presence on fieldtrips in the stated role of researcher. I made cursory fieldnotes as unobtrusively as possible about my observations and conversations because I did not want participants to think that I was taking notes about their every word or movement. At other times it was appropriate to ask, and I did ask, for permission to write down a summary of a particular conversation.

My selection of potential participants for later interviews was based on what I had noticed about individual participants during the participant observation phase. I interviewed students who appeared to be excluded from and/or alienated by the fieldtrip culture. Although I interviewed more students who felt alienated, I also interviewed students who did not experience fieldtrip culture as exclusive and/or alienating. Confirming and disconfirming cases were important in a project informed by an ethos of reconsideration of the construction of geographic knowledge *and* research knowledge.

I handed out post-fieldtrip evaluation forms at the end of fieldtrips or back at the educational institution when the class met for the first time after the fieldtrip had taken place. The questions paralleled the kinds of questions that I planned to ask the interviewees and also included an open-ended question, inviting any kind of comments students might have about their experience of the fieldtrip so that the students could include material that might not have "fit" the other questions. This method offered the advantage of a relatively private and anonymous form of communication in which I might find out if there were other students who felt alienated by the fieldtrip that I had not already noticed. This served as a checkpoint so that I did not base decisions about interviewees solely on observation and/ or overheard comments but also sought students' own written perspectives as an additional source of knowledge.

In a context where students were the ones ostensibly learning how to think like and act like geographers during residential fieldtrips, it was important to ask students about their experiences and memories. I left the interviews as late as possible within the same academic year as the fieldtrip and asked what students remembered based on a notion that memories might be a distillation of their most memorable fieldtrip experiences. "When [students] are acknowledged as experts on their own learning, they articulate very well the connections between life experiences [and] practice . . . conversation uncovers knowledge which may not be evident within other paradigms or structural frames" (Collay, 1989, p. 19). I was concerned with what *seemed* real to each of the students and staff members that I interviewed, rather than with proving what *was* real (if indeed it was possible to do so). I assumed that the interview participants were telling the "truth" about their experiences "insofar as they understood and remembered the events. There was no reason for them to lie, although for various reasons, certain information may have been deliberately left out" (Middleton, 1985, p. 162). This approach is an explicit challenge to forms of knowledge that assume reality can be proved because it is observed and/ or there are a large number of instances. In addition, it constructs learners as experts and memories as a significant source of knowledge.

I conducted two types of interviews as part of the study. The first type comprised post-fieldtrip interviews in which I interviewed participants from the seven fieldtrips some months after the fieldtrips had taken place. The second type included interviews with key informants, people who were not directly connected with the seven fieldtrips, because I also wanted to work at the peripheries of the research topic, to talk "to people no longer actively involved [in fieldtrips], to dissidents and renegades and eccentrics" (Miles and Huberman, 1994, p. 34). I asked particular individuals for an interview because they were prepared to comment on politically sensitive issues (Miles and Huberman, 1994). For example, I interviewed individuals about issues such as the significance of fieldtrips for funding geography as a science in the university context, performances of "alternative" masculinities and femininities, and controversial decisions related to fieldtrip organization. As Miles and Huberman (1994) point out, there are benefits of peripheral sampling in a research project concerned with contradictions and alternative ways of collecting and analyzing data.

I offered research participants the opportunity to be interviewed one-to-one, in pairs, or in small groups of three, four or five. I included the small group (a type of focus group) approach to interviewing because I was interested in the group dynamics of the residential fieldtrip. I was concerned with the interactions between participants as well as with what was said because people's knowledge and attitudes are not entirely encapsulated in reasoned responses to direct questions. Everyday forms of communication such as anecdotes, jokes or loose word association may tell us *as much*, if not *more*, about what people "know." In this sense focus groups " 'reach parts that other methods cannot reach' – revealing dimensions of understanding that often remained untapped by the more one-to-one interview or questionnaire" (Kitzinger, 1994, p. 109).

The interview is an arena of performance – overt and covert – the small group interviews emphasized these aspects of the interview as individual participants performed for their peer group. At times, it seemed that these interviews were not being taken as seriously as the one-to-one or pair interviews were, although (I reasoned) the joking culture of the small group interviews was recognizable as the joking culture of the fieldtrips, recorded in my fieldnotes. In other words, the forms of communication of the small group interview mirrored the forms of communication that I had witnessed on fieldtrips. Ironically, despite my stated interest in other ways of knowing, transcribing and analysis of "irrational" forms of data such as laughter and joking was more difficult. I subsequently favoured interviews with one or two participants because they generated more manageable ("rational") data. Indeed, the academic ritual of research depends on rational rather than irrational data (see Kobayashi and Peake, 1994).

I discovered, however, that interviewing pairs was not always successful. For example, I suggested a pair interview to two female university students who agreed and the interview went ahead. One of the students left for another commitment just before the interview ended. The other student remained on for what turned out to be another quite different interview during which she said things that she would not have said in front of any other student. Similarly, a male student whom I interviewed one-to-one, also acknowledged that he would not have talked about his friendships with male students if he had been interviewed with one of those friends (if that friend/participant had turned up as planned). In other words, pair interviews offered the potential to explore the interactions between participants but could also be constraining in terms of what participants felt able to say. I utilized both approaches (one-to-one and pair) to interviewing, knowing that there were different advantages to be gained from both methods.

Ways of asking were therefore informed by feminist theories and methodologies concerned with how to examine and represent diversity and dilemmas in research. Such a multi-method qualitative approach premised on reconsideration, represents an important challenge to positivist (geography) research practices premised on exploration and discovery.

Ways of Interpreting, or Data Analysis

In this section, I outline my analytical methods, my ways of interpreting. Data analysis was theory-driven: "Choices of informants, episodes, and interactions [were] being driven by a conceptual question, not by a concern for 'representativeness'" (Miles and Huberman, 1994, p. 29).

Students', teachers/lecturers', and researcher's perspectives of residential geography fieldtrips are compared. This is called data triangulation and is one strategy for improving the validity of research findings by exploring what independent data sources "say" about a particular social phenomenon. There are three possible outcomes of triangulation: *convergence*, *inconsistency*, and *contradiction* (Mathison, 1988). *Convergence* of data sources was satisfying because it showed where one or more students, one or more teachers/lecturers, and the researcher independently shared a similar perception of a particular social phenomenon. This improved the validity of the research findings. Nevertheless, the *inconsistencies* and *contradictions* between information from different data sources were just as important because they directed my attention to what had gone unnoticed or had been perceived differently by students, staff, and the researcher. *Inconsistencies* and *contradictions* between information from different data sources can push researchers to reformulate previously taken-for-granted understandings and potentially extend explanation of social phenomena. In a feminist project contesting the reproduction of (geographic) knowledge, such provision for analysis and inclusion of contradictory evidence is important.

Retrospectively I knew I had more data than I needed about high school fieldtrips. This over-compensation was in part due to the politics of the field. I found it much easier to gain entry to high school fieldtrips and later to interviews with high school students, than I did to university fieldtrips and interviews with university staff and students. My concerns about having enough data were also predicated on a theoretical goal of remaining open to unexpected as well as expected themes. I reasoned that taking notes about what seemed irrelevant at the time could prove to be relevant to emerging and/or unexpected themes in later fieldtrips. Determining how much data are needed in advance (rather than retrospectively) might be achieved by maintaining data analysis concurrent with collection. An ongoing assessment is then possible of which research questions have enough data available so that selected forms of verification, such as triangulation, are meaningful. And in a parallel fashion, it is possible to identify the gaps in data which mean that specific research questions cannot be meaningfully answered until more data is collected.

These suggestions imply a straightforward pragmatic approach that belies the politics of conducting feminist qualitative research. The challenges to the legitimacy of the topic referred to at the beginning of the chapter remained with me throughout the project. I therefore collected more data because I myself was inculcated in disciplinary forms of data collection informed by positivist and quantitative paradigms where more data symbolized more proof. More proof, although illusory and debatable, was seductive because I wanted my feminist research to be taken seriously by colleagues. In hindsight, I realize that no matter how large the quantity of data, the legitimacy of feminist research will continue to be challenged because it contests the very basis of knowledge claims.

In my analyses of the diverse and large quantity of data collected during this research, I primarily utilized qualitative methods but employed quantitative methods to summarize, support and/or test the findings that I arrived at via qualitative approaches. I discuss my analysis of the drawings as one example of how I utilized three ways of interpreting the drawings rather than rely on one approach. First, a quantitative approach in which I counted activities and environments then tried different classification sys-

Figure 8.1   Female student's drawing of what she expects to do on a field trip

Draw what you think geography student(s) do on a fieldtrip . . .

tems such as formal, informal (or both) for activities, and indoors, outdoors (or both) for environments to test general ideas about students' expectations of particular environments and activities on a forthcoming fieldtrip. Second, a poststructuralist analysis in which I examined the implicit binaries represented in the drawings such as outdoors/indoors, night/day, student/staff, and work/fun. I then analyzed particular drawings where such binaries were implicit and explicit, to find out if such "readings" supported emerging themes or not, and selected exemplary drawings. In the case of one drawing this binary was explicit; the page was divided clearly between night-time and day-time activities (see figure 8.1).

The third mode of analysis was to consider the drawings as representing "an embodied social world" (Du Plessis and Fougere, 1995, p. 132). Drawings in particular relied on the use of "objects imbued with symbolic significance" (Du Plessis and Fougere, 1995, p. 132) such as "the mortar board" to indicate status, "the tree" to indicate the outdoors, "the beer can" to indicate social activities, to tell a larger story (a picture is worth a thousand words!). I had asked students to draw pictures to undermine the

privileging of words in the academic arena (see Monk, 1997). It was ironic
that I re-presented these drawings in the words of analysis. (Perhaps I
should have been drawing conclusionary pictures?) I countered the representation of drawings in words by including as many drawings as
possible in my dissertation for the reader to "add" their analysis to mine.
But this goal was confounded by the "quality" of some drawings, that is,
those drawings of a quality easily scanned were more likely to be included
in the dissertation.

The analysis of data generated during participant observation and the
post-fieldtrip interviews involved additional analytical strategies, namely
the writing of theoretical, methodological and analytical memos (see Glaser, 1978), and the organization of data around emerging themes. "A
memo can be a sentence, a paragraph or a few pages. It does not matter as
long as it exhausts the analyst's momentary ideation based on data with
perhaps a little conceptual elaboration" (Glaser, 1978, p. 84). "I wrote the
memos for myself and they tie[d] together different pieces of data into a
recognizable cluster, often to show that those data [were] instances of a
general concept. Memos can also go well beyond codes and their relationships to any aspect of the study – personal, methodological, and substantive. They are one of the most useful and powerful sense-making tools at
hand" (Miles and Huberman, 1994, p. 72).

My theoretical memos included comments about emerging themes,
contradictions, and notes on new ideas that might explain puzzling data
(Miles and Huberman, 1994). My methodological memos included notes
about changing methods of data collection and "doubts about the quality
of some of the data," and "ethical dilemmas" (Miles and Huberman, 1994,
p. 66). Analytical memos included preliminary summaries of data, even
preliminary conclusions that acted as place-holding analyses requiring
further data, as well as "cross-allusions to material in another part of the
data set," and "elaboration or clarification of a prior incident or event that
seem[ed] of possible significance" (Miles and Huberman, 1994, p. 66).

During data collection and analysis, four recurring themes emerged: (1)
the difference between the place(s) visited on geography fieldtrips and the
place(s) where geography students live and attend their place of education;
(2) the significance of the fieldtrip as an opportunity for going out in the
"real" world and "seeing" for yourself; (3) the experiences of living and
working together on a residential fieldtrip; and (4) the "re-creation" of
geographers during fieldtrips.

I created separate files around each of these four themes. I will explain
how I did this in relation to the second theme, the significance of the
fieldtrip as an opportunity for going out in the "real" world and seeing for
yourself, as an exemplar of my approach. I did a keyword search of all the
summarized transcripts for words such as "real" and "see," and placed all

of the relevant sections of text with associated details such as name of student and fieldtrip, into separate files for each of the four institutions. I then read all four files, highlighting key quotes, writing marginal notes, cross-referencing, and continuing the process of writing memos. Marginal notes were an important way of recording "new interpretations, leads, connections with other parts of the data, and they usually pointed towards questions and issues to look into . . . and to ways of elaborating these ideas" (Miles and Huberman, 1994, p. 67).

For each theme, I searched for inconsistencies and contradictions in four ways. First, I purposefully re-read the files generated around each theme for inconsistencies and contradictions. Second, I did further keyword searches for contradictory words. For example, I searched for words indicating senses other than seeing (hearing, tasting, touching, smelling), in particular for the words – "hands-on" – in the case of the second theme. Third, I searched for metaphor and analogy (Bogdan and Biklen, 1992) which often revealed inconsistencies and contradictions in the privileging of particular terms in geographical discourses. For example, I searched specifically for instances where the word seeing was deployed metaphorically but did not make sense literally. Fourth, I identified and analyzed binaries because contradictions and inconsistencies were often inherent in particular binaries. In the case of the second theme, I returned to the sections of text under the headings of "real world/textbook" and "theory/practice" in the summarized transcripts because these binaries were evident in much of the participants' talk about going out to see the "real" world. I even engaged in extended discussions about these binaries (without naming them as such) with some interviewees, in response to their claims about the effectiveness of fieldtrip learning compared to classroom or lecture theater learning. Evans (1988, p. 214) also describes how he tried out emerging theories and hypotheses during conversations within the research situation. I then analyzed these extended discussions closely for congruencies, inconsistencies and contradictions. The final act of analysis was the process of writing itself.

The analytic induction of categories, themes and relationships; the explication of meaning; and the understanding of action may all proceed via the writing itself . . . The "writing up" of the qualitative study is not merely a major and lengthy task; it is *intrinsic* to the "analysis," the "theory," and the "findings" (Atkinson, 1991, p. 164; emphasis in original).

Reconsiderations, or Drawing Conclusions

To conclude is to reconsider. In line with my stated goals of challenging the construction and reproduction of geographic knowledge, drawing conclusions is a form of reconsideration and re-interrogation. In writing conclusions, it is "of course, always a positing, and hence excludes and demarks, thus always itself open to the possibility of deconstructive technique" (Young, 1990, p. 321). At the end of a project in which I was critical of certainty in (geographic) knowledge claims, it was difficult to write conclusions with any finality or certainty.

To critique and to deconstruct is a well-traversed academic path. Ways of asking and ways of interpreting described in this chapter are primarily deconstructive. But it is not enough to simply deconstruct and to destabilize (see Young, 1990). Any research informed by a feminist notion that a different gendered "reality" is possible requires more than this. The existence of instabilities in the binary logic of sexual difference has no necessary effect on the politics of sexual difference (Sedgwick, 1990). A further step must be taken if feminist or gay or anti-racist struggles are to benefit. In other words, "some kind of *social practice must lean on these instabilities* if they are to represent any kind of transformative possibility" (Waldby, 1995, p. 274; emphasis added).

In feminist poststructuralist analysis, one term implies its opposite. Within the findings that emerged from deconstructive ways of asking and interpreting, alternative social and epistemological ways of conducting fieldwork and fieldtrips were implied/suggested. For example, a deconstruction of the privileging of the visual simultaneously models how the basis of knowledge claims can be challenged and suggests other forms of knowing because the visual is called into question.

Multi-method data collection and analysis informed by feminist politics is a means of "acquiring and codifying knowledge" (Nagar, 1997, p. 203) that goes "against the grain" of dominant masculinist ways of (geographic) knowing. But it is not a simple matter of pitting feminist against masculinist ways of knowing. The (re)production of feminist knowledge must also be subjected to critical reconsideration. A multi-method approach increases the likelihood that contradictory (feminist *and* masculinist) knowledge claims might surface and be more thoroughly investigated.

I utilize the moments in which I make summaries of my research findings to argue for strategic interventions in the social and epistemological practices of residential fieldtrips that lean on the instabilities and tensions exposed theoretically and methodologically. To "re-construct," or in the terms of this project, to posit alternative geography fieldtrip practices, is risky because it entails a misunderstanding of these alternatives as prescriptive and/or as answers to all the identified issues. In spite of these risks, it is important to move beyond critical analysis to reconsideration of what could be different if geography fieldtrips were not so central to the geography discipline and/or were conducted in unconventional ways. In turn, such interventions and alternatives must be continually subjected to critical re-evaluation (see Alton-Lee and Densem, 1992). Theoretical and methodological maneuvers need to affect social and epistemological practices, and be critically reconsidered, if the feminist project of disrupting masculinist forms of knowledge is to benefit women and men.

ACKNOWLEDGEMENTS

I would like to thank the editor of this collection, Pamela Moss, and colleague Ruth Liepins for their encouragement and helpful comments on earlier drafts.

RESEARCH TIP

What to Put in Your Interview Bag

- Tape recorder
- Tapes/back-up tapes
- Batteries
- Pens
- Elastic bands
- Tissues
- Pertinent information, e.g. address, phone number, map with directions
- Business card or equivalent
- Brief, jargon-free written project description
- Letter of informed consent
- Documentation of ethics approval
- Interview guide
- Research journal
- Phone numbers for counseling (if appropriate)
- Gift (if appropriate)

9

Quantitative Methods and Feminist Geographic Research

Mei-Po Kwan

Introduction: Quantitative Geographical Methods

Quantitative methods not only involve the use of numbers such as official statistics. They include the entire process in which data are collected, assembled, turned into numbers (coded), and analyzed using mathematical or statistical means. In research relying mainly on quantitative methods, the focus of the data collection effort is to gather quantitative data or qualitative information that can be quantified in some way (as in attitudinal studies). Once coded, these data are then explored using various methods, ranging from simple measures such as frequency counts and percentages, to complex techniques such as log-linear models. Results of quantitative analysis can be presented in the form of summary statistics, test statistics, statistical tables, and graphs. They can also be represented in complex cartographic or three-dimensional forms with the assistance of GIS, or Geographical Information Systems.

Since spatial data violate many assumptions of conventional statistical methods, such as independence of each individual observation and constant variance, quantitative analysis of geographic data calls for spatial statistical methods that were developed to overcome the problems of applying conventional statistical methods to geographic data. This is a specialized collection of techniques required when dealing with data that describe the spatial distribution of social or economic phenomena, many of which are of interest to feminist geographers. Without applying the appropriate geostatistical methods, analysis of geographical data may lead to erroneous results and conclusions. There are other new developments in quantitative geographical methods which are particularly relevant to feminist research. The recent development of local statistics facilitates the analysis of the relationships between the local context and women's everyday experiences.

Recent use of GIS-based geocomputational and geovisualization methods represents another area with potential for feminist research (Kwan, 2000a).

Quantitative methods have been used in feminist geographic research since the early days of feminist geography. The original intention was to produce a more accurate and less "biased" description of the world by studying the world through women's perspectives and experiences (Moss, 1995a). Recent debate in feminist methodology provides helpful insights for the use of quantitative methods in feminist research (Mattingly and Falconer Al-Hindi, 1995). As a feminist geographer, it is important to understand the limitations and value of quantitative methods, when quantitative methods are appropriate, and how to approach using them. In this chapter, I examine critical issues concerning the use of quantitative methods in feminist geographic research and illustrate some of the steps for undertaking quantitative feminist research using my recent work as an example.

Feminist Critiques of Quantitative Methods

Quantitative geographical methods were developed during a period now commonly called the "quantitative revolution" in geography. These methods were developed with the intention of making geography a scientific discipline not unlike physics, where the validity of the knowledge was justified according to positivist principles. With a positivist epistemology, the purpose of geographic research was to seek universally applicable generalizations. The researcher was considered a detached observer capable of acquiring objective knowledge of the world through discovering empirical regularity in social, economic or spatial phenomena.

Early feminist geographic research emerged to show the neglect of women's experiences and to include women as subjects in geographic research using largely quantitative methods (McDowell, 1993a). As questions about women were added to geographical inquiry using methods similar to quantitative geography, early feminist geographic research was considered positivist and empiricist because it was based upon the principles of scientific objectivity, value neutrality, and the search for universally applicable generalizations. Feminist geographers who did these kinds of studies, as feminist critics argued, intended to make geography a better "science" through correcting male bias and using more stringent scientific methods. Their work was considered empiricist as they privileged claims to knowledge based primarily upon observable "facts." Feminist critics also asserted that "truths" put forth as universally applicable are valid only for men of a particular culture, class, or race (WGSG, 1997). They are also critical of the tendency to derive analyses of universal causality from inferential statistics.

Quantitative methods were criticized by feminists for other reasons. For instance, since quantitative methods depend on some quantifiable attributes of the phenomena under study, they are not capable of reflecting the complexity and richness of women's lives. This is a serious limitation since a substantial portion of women's experiences cannot be expressed by numbers and is therefore not quantifiable. Further, the "live connections" with research subjects are often lost through the use of quantitative data, making it difficult to tell women's feelings and their interactions with others. This in turn makes it difficult to obtain a contextualized and holistic understanding of the complex processes involved in determining their everyday experiences. Quantitative data and methods are therefore "disembodied" – as abstracted and decontextualized information is used in the process (WGSG, 1997).

Feminists also criticized the assignment of any specific individual's experience into hard-and-fast categories in the collection and analysis of quantitative data, whether these categories are predefined by the researcher or according to official criteria (Jayaratne and Stewart, 1991). The rigid nature of the categories and variables used may fail to reflect the complexities of women's lived experiences. Very different phenomena may be lumped together in the statistics as if they were the same thing and the statistics may have a problematic connection with the life they claim to represent (Pugh, 1990). Further, since pre-existing categories and official statistics were often based on male experiences, using them in feminist research can be self-defeating. They may actually make it difficult, if not impossible, to reveal the processes underlying the inequality and oppression women experience (see for example Perrons, 1999). For instance, official statistics were often found to be unreliable and even useless for studying women's labor force participation or contribution to the economy because many forms of women's unpaid work are omitted in official definitions of "work" (Samarasinghe, 1997). Another example is Pugh's (1990, p. 107) study on homelessness, where she concluded that "life will always be more complex and ambiguous than any possible usable system of coding and classification."

Approaching Quantitative Methods as a Feminist Geographer

Epistemological considerations

So, how do feminist geographers go about using quantitative methods in their research? The first thing is to identify alternative, critical practices of quantitative methods that can, at least to a certain extent, address the concerns of feminist critics. Insights from recent debate in feminist methodology are particularly helpful here. One is that using numbers and quantitative methods is not the same as holding what sort of knowledge is valid or privileging certain kinds of knowledge over the others (Lawson, 1995; McLafferty, 1995). The association between quantitative methods and positivist/empiricist geographic research was more historical than necessary or unchangeable. Feminist geographers need to move beyond the kind of scientific objectivity and value-neutrality that characterize quantitative geography of earlier periods.

As feminists now hold, the kind of scientific objectivity that is based upon the existence of a detached, transcendent observer is not only unachievable but masculinist (G. Rose, 1993). Feminist objectivity should be understood in terms of the situated knowledges based on particular "standpoints" or limited "positions" of women's lived experiences in particular social and geographical contexts (Haraway, 1991; Harding, 1991). Further, the use of quantitative methods by itself does not confer the researcher any authority to make privileged knowledge claims as compared to other forms of knowledge, especially those obtained through qualitative methods. Rather, it helps to situate other forms of knowledge in the context of the overarching social and economic relations (Moss, 1995a). Feminist geographers using quantitative methods should limit their conclusions rather than making grand claims about the universal applicability of their results (Rose, 1997).

Beyond the qualitative/quantitative dualism

Another important point is that criticisms of quantitative methods, as a reaction to positivism and empiricism, can lead to an unuseful oppositional stance that holds qualitative methods as the preferable alternative to quantitative methods (Harding, 1987a). This, however, not only perpetuates dualist thinking through holding a qualitative/quantitative dualism that characterizes masculinist thinking, but also ignores the possibility of postpositivist, critical quantitative methods that are consistent with feminist epistemologies and politics (Lawson, 1995; Sheppard, 2001; Sprague and Zimmerman, 1989). It is perhaps more helpful to think of quantitative methods as one of many possible feminist methods that can be used together with other methods. As the analysis of quantitative data can be complemented by a contextualized understanding of women's everyday lives provided by qualitative data, and the interpretation of qualitative data can be assisted by the broad picture provided by quantitative methods, using multiple methods in a single study may provide a more complete understanding of the questions at hand. This strategy of "triangulation" has advantage because the weaknesses of each single method may be compensated by the counter-balancing strengths of another (D. Rose, 1993).

In the practice of feminist research, it is important to recognize the limitations and strengths of quantitative methods. Quantitative methods simply cannot provide the kind of rich and contextualized account of women's experiences that qualitative methods can permit (Jayarante and Stewart, 1991). They are therefore more suitable for answering certain questions and are less appropriate for addressing others. Feminist researchers need to determine the appropriateness of quantitative methods and their combined use with other methods for a given research question. It is also important to identify the research question based on critical feminist concerns and/or feminist theories before deciding which method(s) one will use in a particular study. The primary issue is what data are needed and which methods are appropriate for addressing the research question.

Data problems

There are other concerns for the practice of quantitative methods in feminist geographic research. Quantitative data can come from secondary data sources, such as official statistics. They can also come from primary data collection through surveys. Since the counting procedures or classification schemes used to collect official data often ignore significant aspects of women's lives and experiences (e.g. counting women's work and male violence against women), the collection of primary quantitative data is a better strategy than the reliance on official statistics for many issues of interest to feminist geographers. A good example is a study discussed in Reinharz (1992, p. 82) by two law students who collected data from a judge and police chiefs to show the prevalence of wife battering in the local area.

Another issue is that great care is needed when developing a coding scheme because rigid categorization is a major weakness of quantitative methods. For example, social differentiation should be defined by using many dimensions, such as gender, race, ethnicity, class, and sexuality. The use of more refined coding schemes for classifying individuals into social groups would yield better understanding of significant differences between individuals than one based upon any single criterion such as gender. The use of advanced categorical data analysis techniques that can consider several differentiating dimensions at the same time is also preferable to those that are based on a single dimension at a time. Further, presentation of quantitative data should be accompanied by a description of the ambiguity or problems of the classification scheme. Any reservations about the results because of this should also be provided. An evaluation of the sensitivity of the results to different classification schemes will be even more helpful to the audience.

Measurement issues and statistical analysis

Another important issue in quantitative methods concerns the quantification process and the analysis of quantitative data. Before quantitative data can be collected, concepts central to the research question need to be *operationalized*. This means that the researcher has to determine how various phenomena are to be measured and how the required data are to be collected. Turning concepts such as "class" or "discrimination" into quantifiable measures is far from straightforward. Feminist geographers therefore need to deal with all operational issues with care. For example, how should one measure women's "household responsibility"? One commonly used measure is the number of children in the household, which is unlikely to be a good measure because it may not have a consistent relationship with the amount and type of domestic tasks women perform.

As conventional geographical concepts and existing quantitative measures may contain serious male bias, the question about the gender sensitivity of these concepts or measures is also important. It is important for feminist geographers to critically re-assess all existing measures and look for any such bias before using them. It may be necessary to develop one's own method of counting or measurement for the research question (see for example Kwan, 1999b).

It is important to note that some feminists argued against the use of inferential statistics in feminist research, where only non-parametric and descriptive statistics are considered appropriate. Feminist geographers therefore also need to understand the concepts of statistical inference and significance, and to situate these techniques in the context of feminist epistemologies. Although all statistical inferences, including non-parametric statistics, assume some notion of "typicality" in circumscribed populations, using inferential statistics does not necessarily mean making totalizing generalizations or asserting universal causality (Pratt, 1989). Inferential statistics are based upon our understanding of the likelihood of occurrence of certain events. They can provide a basis to determine whether the phenomena observed is typical or not for the population subgroups being studied (without arguing that the relationship observed is also true for the larger population). If there are wide variations in what individuals experienced in a sample, it is difficult to argue that it is shared by members of the group. If differences among various subgroups of individuals are statistically significant, such differences are unlikely to be caused by chance alone and therefore deserve a closer look. Inferential statistics can therefore be used in feminist research in a non-generalizing, non-totalizing manner.

The Place of Quantitative Methods in Feminist Geographic Research

Since the strengths of quantitative methods are in describing and analyzing complex patterns of social, economic and geographical phenomena of interest to feminist geographers, they can be used for certain purposes. First, quantitative methods are useful for describing the measurable aspects of women's everyday experiences and analyzing complex spatial relations among geographical phenomena. They are particularly helpful for providing a broad "picture" of the social, spatial or temporal inequalities women experienced at various spatial scales. As McLafferty (1995, p. 438) argued, quantitative methods can reveal "the broad contours of difference and similarity that vary not only with gender but also with race, ethnicity, class and place." Quantitative methods are therefore especially valuable when there is an urgent need to have a broad "view" of women's current situations (e.g. male violence against women), but detailed individual-level data are not readily available or the limited resources at hand prevent the collection of qualitative data. They also help to highlight the shared experiences of many similarly situated women such as domestic violence and sexual discrimination (Moss, 1995a).

A good example is the research by McLafferty and Preston (1992) which used aggregate census data to analyze the complex relationships among gender, race, ethnicity, occupational status and commuting distance. Their studies showed that the well-known gender differences in the length of the commute trip varied considerably among race and ethnic groups. Although the categories employed in their quantitative analysis, like gender or ethnicity, can be questioned, and the data they used did not give a contextualized understanding of the lives of the people they studied, their research indicated that quantitative methods can still be used to describe and analyze the similarities and differences among groups of women at different times and places.

Another way in which quantitative methods are useful in feminist geography is that the presentation of quantitative data or results of quantitative analysis is often more forceful in political discourse than the use of qualitative data. This is especially true as "hard" data obtained using quantitative techniques often appear to be more convincing to public policy makers. Surveys may have the power to change public opinion in ways that a limited number of in-depth interviews may not. For instance, Seager and Olson (1986) documented the extent to which women were unequal and subordinate to men throughout the world using official statistics. They concluded that women everywhere are worse off than men – they have less autonomy, less power, less money, but more work and responsibility. Several studies had already shown that women in general have more spatially restricted lives than men – they work closer to home and travel less – and are often employed in female-dominated occupations and earn less than men (see for example Tivers, 1985; Hanson and Pratt, 1995). As Lawson (1995) argued, descriptive data like these powerfully present the unequal power and gender relations within the household and the economy at large. In describing certain measurable aspects of women's lives, descriptive data reveal the social and political processes that help to perpetuate the inequality and oppression of women.

In light of this, quantitative data and methods may be a powerful instrument for initiating progressive social and political change. They may help reduce the marginalization or oppression of women. For example, using a large survey data set, Rosenbloom and Burns (1994) documented that working mothers rely heavily on the car to balance their domestic and child-care obligations. Travel demand management measures that aim at reducing travel without taking their needs into account will have considerable negative impacts on their lives. This result can be used to steer public policies to better meet the need of working mothers. Another example, cited in Jayarante (1983), is the court decision of a sex discrimination lawsuit that began to make statistics acceptable as legal evidence.

Closely related to this is that the analysis of quantitative data may stimulate questions about the process of oppression or gender relations that generate the numbers. This may help reveal research areas that urgently require attention and indicate directions for more in-depth and qualitatively oriented research. For example, in a study by Tempalski and McLafferty (reported in McLafferty, 1995), quantitative analysis helped identify the lower-middle income neighborhoods in New York City where the problem of low birthweight is serious. With these results, healthcare and social work professionals can undertake in-depth qualitative research in these areas to obtain a better understanding of the problem. Quantitative methods can also be used to reveal and challenge the male bias in existing geographical concepts and methods. For example, in my research on conventional measures of accessibility, I found that all conventional accessibility measures failed to take women's needs to undertake multipurpose trips and their space–time constraints into account and therefore suffered from a serious male bias. This led me to formulate and implement space–time measures of individual accessibility that can better reflect women's individual access to urban opportunities (Kwan, 1998, 1999b).

An Example: Gender, Work and Space–Time Constraints

The example described below is from my recent research on the complex relationships between women's commuting distance, employment status, and space–time constraints. Detailed theoretical arguments, methods, and results are elaborated in separate publications (Kwan, 1999a, 1999b, 2000b). The project was built upon earlier studies that include many excellent examples of using quantitative methods in feminist geographic research (see for example England, 1993; Hanson and Pratt, 1990, 1995; Johnston-Anumonwo, 1995, 1997; McLafferty and Preston, 1992, 1996, 1997).

Research question

I formulated my research question in light of two recent trends. First, as more and more women participate in the labor force, some of them have been able to achieve relatively high occupational status and income. Second, as the proportion of women who can use their own automobile to commute increases, many women now have much better spatial mobility than before. Many believe that these two trends together will lead to changes in the allocation of domestic responsibilities within the household (at least for those women who have achieved high occupational status and spatial mobility). If this is true, it also means that men will take up a large proportion of household responsibilities. These changes in the domestic division of labor in turn will hopefully be associated with changes in the gender relations within the household. My study attempted to find out whether these trends actually lead to changes in the allocation of household responsibilities. I also sought to examine whether the constraints associated with women's need to perform domestic responsibilities are still important in determining their employment status and commuting distance.

The concepts used in the study are based upon earlier research on the geographies of women's everyday lives using time-geographic concepts (see for example Tivers, 1985). The time-geographic concept most relevant to feminist research is "space–time constraints," which impact upon women's daily lives in significant ways and stem from two main sources. First is the limited time available for a person to perform various activities within a particular day – commonly referred to as time budget constraint. The second source, referred to as fixity constraint, arises from the fact that activities that need to be performed at fixed location or time (e.g. child-care drop-off) restrict what a person can do for the rest of the day.

Past studies observed that space–time constraints significantly affect women's job location, occupational status, and activity patterns. One limitation of this literature is that none of these studies attempted to measure fixity constraint directly and assess the extent to which it impacts upon women's employment status and commuting distance. My research attempted to address this limitation through collecting quantitative data about the spatial and temporal characteristics of individuals' activities and analyzing their relationships with women's household responsibilities, job location and employment status.

Formulation of operational measures

Before setting out to collect the data, I had to resolve operational issues about how to turn the notion of space–time constraints into something measurable. Based upon previous work on this area, I decided to solicit information about the space–time fixity of each activity a person performed through an activity–travel diary survey. The diary recorded details of all activities and trips made by the respondent in two designated travel days. I included four specially designed questions in the diary to obtain information about the spatial and temporal fixity of each activity (see Kwan, 2000b). Using answers to these four questions, I designated three types of fixity: (a) spatially fixed activities; (b) temporally fixed activities; and (c) activities which are both spatially and temporally fixed.

Another operational issue involved identifying the purpose of each activity performed by the respondent. A common approach in past studies comprised categorizing activity purposes and then coding the written description by the respondent. One major difficulty of this approach is that the primary purpose of an activity may not be reflected from the written description of the activity given by the respondent. For example, an activity can be performed for different purposes by the same person (e.g. grocery shopping may be undertaken for meeting household need or for social or recreational purposes), and the same activity may be performed for different purposes by different individuals. To overcome this problem, I included a question in the activity diary to record the primary and secondary purpose of an activity according to the respondent's subjective evaluation. Five activity purposes were initially provided to the respondent as guidelines, but they can also provide their own answers in an open-ended question.

Data collection

After resolving operational issues, I developed the survey instrument. It includes two main parts: a household questionnaire and a two-day activity–travel diary. Using this survey instrument, I collected data from a sample of adults (over eighteen years of age) in households with one or more employed member(s) in Franklin County, Ohio, in 1995. The household questionnaire collected information about the socioeconomic characteristics and transport resources of all household members. The two-day activity–travel diary collected detailed information about the activities and trips of the respondent for two designated days. Data collected included street address, travel mode used, car availability, routes taken, the primary purpose of each activity, a subjective fixity rating for each activity, and other individuals present when performing each activity.

Because the small number of ethnic minorities in the sample does not allow for meaningful statistical analysis, they were excluded from the analysis (this would not have been the case if qualitative information had been collected). The final subsample consists of three groups of European Americans (white): twenty-eight full-time employed female, thirteen part-time employed female (who work less than thirty-five hours a week), and thirty-one full-time employed male.

Analysis and results

I analyzed the differences in fixity constraint experienced by individuals of these three groups using simple descriptive statistics and analysis of variance (Kwan, 2000b). The results show that women employed part-time encounter more fixed activities in their daily lives than the other two groups. Many of these fixed activities are associated with household needs that have a strong restrictive effect on the locations of their out-of-home activities and job location. Further, despite the fact that women employed full-time travel longer to work than men, they experience higher level of fixity constraint than men. This result is surprising considering the high occupational status and high level of access to private cars of the full-time employment women in the subsample.

The results of a canonical correlation analysis I performed reveal that, for individuals in the subsample, the level of day-time fixity constraint depends more on one's gender and the extent to which household responsibilities are shared with other adults in the household, than on some conventional variables of household responsibilities such as the presence or number of children in the household (Kwan, 1999a). To analyze the complex interrelations among women's day-time fixity constraint, non-employment activities, household responsibilities and employment status, I estimated a nonrecursive structural equation model with latent variables for the women in the subsample (Kwan, 1999a). The results show that fixity constraint has a significant impact on women's employment status (where women with higher levels of fixity constraint are more likely to work part-time).

Overall, these findings suggest that, women in the subsample face higher levels of fixity constraint than men in the subsample, regardless of the length of their commute trips and their employment status. The experiences of these women therefore allow us to question the belief that increasing female participation in the labor force will lead to significant change in women's gender role and space–time constraints. The results also suggest that the situation of women may not change much without first changing the gender relations and redressing the division of domestic labor within the household. Despite the belief that recent trends in the increasing number of women with higher occupational status and improvement in their access to private means of transportation will lead to changes in traditional gender roles, the results of my study call into question such a belief.

Presentation of results

There are several qualifications applicable to these results. First, given the specific subsample and context of the study, its results cannot be generalized to other gender/ethnic subgroups or sociospatial contexts. For instance, given the high socioeconomic status and travel mobility of the individuals in the subsample, the results may seriously understate the fixity constraint and mobility problems faced by individuals of other gender/ethnic subgroups (especially minority women). Further, the survey data used in the study do not allow for the examination of other important factors such as labor market processes and the negotiation between the female and male heads of household. The interaction between these factors and women's space–time constraint is an important issue for future research. In view of these limitations, I realized that complementing the results with ethnographic data of the individuals could have led to a better understanding of the complex processes involves (for example, women's fear of violent crime may impose significant space–time constraint on their activity patterns).

Using Quantitative Methods as a Feminist Geographer

My study focused on an important aspect of women's everyday lives that also partly reflects gender relations within the household. I developed and used measures which I considered more appropriate and more capable of reflecting the complexities of women's daily lives. I collected individual-level data from a sample instead of using secondary data, thus avoiding the many omissions one may encounter when using government surveys. I used advanced statistical techniques to analyze the complex interrelations among women's domestic responsibilities, occupational status and space–time constraints. In the study, I also developed GIS-based computational and geovisualization methods for exploring the data without first reducing the original data to statistical aggregates, thus retaining the particularities of each individual subject. To make the interpretation of the results less disembodied, I also talked to some of the subjects to clarify issues over the phone, which gave many insights into how to represent these details in the data.

Given that one purpose of feminist geography is to improve our under-standing of the gendered nature of social life and to provide knowledge useful to the struggle for gender equality, quantitative methods can play a role in feminist geographic research. When using these methods as a feminist geographer, special attention has to be paid to epistemological issues. In addition, as misuses of quantitative methods can lead to errone-ous and misleading findings, it is important to understand the proper procedures for undertaking quantitative geographical analysis and what conclusions the data or method allow. This knowledge would also be useful for a feminist geographer to identify the masculinist bias in existing quantitative data and methods.

RESEARCH TIP

Computer Software

Computer software can assist with data management, coding, and analysis. Take the time to make sure that the program you choose is appropriate for your research needs.

- Data management programs, e.g. Excel, Lotus, QuattroPro, Access.
- Quantitative Analysis Packages, e.g. SAS, SPSS, MiniTab, MYSTAT.
- Qualitative Analysis Packages, e.g. Ethnograph, NU*DIST, Text-based Beta.
- New Generation Interactive Qualitative Analysis Packages, e.g. NVivo.
- Bibliographic Referencing Systems, e.g. EndNote, Reference Maker.

There are competing reviews of software analysis packages. We found Renata Tesch's (1991) article useful.

10

Borderlands and Feminist Ethnography

Joan Marshall

The men on this island are living in a time warp from the fifties. (Thirty-seven year-old woman on Grand Manan Island, June, 1999)

Four years ago, during the early stages of my research on Grand Manan, someone described the women there as "the strong women of Grand Manan." This description continues to haunt me, even as my own under-standings and interpretations have evolved. These "strong" women are both complex reflections and creators of a society that is historically rooted in notions of community, family, and the core myth of rurality. Grand Manan women are survivors whose identities are embedded in a long history of rural isolation and resource dependence, and who they are cannot be separated from their historical, geographical and economic contexts.

In opting for a feminist ethnographic methodology as the means to ensure the nuanced research that Chouinard (1997) talks about that would hopefully fill a gap in geographic research, I was guided by the belief that an understanding of women's roles and relationships on Grand Manan would be possible *only* by exploring their self-defined understandings, meanings, and experiences. Their own words, stories, and self-described patterns of behavior would be crucial to achieving my research objective, that is, to understand the evolving processes of local–global interactions in terms of impacts upon women and community. Yet in its intrinsically personal interactive *modus operandi* and its philosophical underpinnings, ethnography is more than a methodology. "The ethnographic stance is as much an intellectual (and moral) positionality, a constructive and interpre-tive mode, as it is a bodily process in space and time" (Ortner, 1995, p. 173). Engaging feminist ethnography as a holistic process necessarily involves ambiguity and fluidity. The feeling of inhabiting *borderlands* throughout this research project has informed my approach, infusing ethnographic research with a sense of living and working in unstable worlds of meanings and perceptions. Using the idea of a borderland, rather than boundary, underlines the fluidity and constantly negotiated realms of being that describe the research process in community settings (see Marcus, 1998; Katz, 1994).

Three Borderlands in Ethnographic Research

At this historical moment and in all the geographical sites of research, it is crucial that social scientists inhabit a difficult and inherently unstable *space of betweenness* in order to engage in rhetorical, empirical, and strategic *displacements* that merge our scholarship with a clear politics that works against the forces of oppression. (Katz, 1994, p. 67)

For the community and for the researcher alike, Grand Manan is both a nexus (meeting place) and a borderland (unstable world of interpretation and meaning), existing between two cultures, informed by the past but profoundly influenced by new technologies and the consumerism of modernity. From the beginning of the project, it became clear that Donna Haraway's "situated knowledges" (1988) defined both methodology and epistemology in ways that necessarily focused less on a sense of a concrete "field" and more on interlocking, multiple sociopolitical sites and locations (Gupta and Ferguson, 1997). As Haraway (1988, p. 590) points out, "Situated knowledges are about communities, not about individuals. The only way to find a larger vision is to be somewhere in particular."

Even in my initial choice of field sites, there was a sense of inhabiting a borderland in my experience of both the familiar and the strange. Borderlands were embedded in my unquestioned assumptions of difference because I felt that "the edges of society" engaged my interest "as challenges to the apparently overwhelming homogenization and hegemony of Western modernity" (Nadel-Klein, 1997, p. 97). Three especially relevant borderlands permeated the entire research process for me: the common ethnographic issue of insider–outsider status, the presence or absence of a feminist perspective in understanding gender relations, and the space between research and action.

The dilemma of determining *insider–outsider status* has many dimensions of concern. As part of a borderland, it defines both the methodology of the research process and the communication of the resulting meanings, stories, and interpretations. The notion of insider–outsider relates, for example, to spatial practice, insofar as researchers move "in" and "out" of their field sites. Ethnography is more than just "passing through"; it demands commitment in time and spatial engagement that embodies "a flexible range of activities, from co-residence to various forms of collaboration and advocacy" (Clifford, 1997, p. 191). As I shall describe, the underlying tension inhabiting the regions between researcher and friend, knowing that visits are timed and that over the long term these relations will evolve more securely toward one pole or the other, inevitably surfaces in every encounter. Clifford's description of the field site as a "habitus" characterized by "a cluster of embodied dispositions and practices" (Clifford, 1997, p. 199) is helpful in understanding the practice of feminist ethnography as a dynamic project.

Intrinsic to the insider–outsider dilemma is the concept of positionality as a perpetual source of questioning and self-revelation (Dyck, 1993). Although initially we may see ourselves primarily as researcher, other roles, past experiences, and our own personal subjectivities significantly define how we conduct our research (Robinson, 1994). Not only do past experiences such as advocate, volunteer, wife, mother, and church member, impact upon how we position ourselves, but evolving aspects of positionality with respect to friend and researcher, locals and visitors, continuously affect how we participate in daily activities and respond to community norms. In an earlier article, I describe the central importance of the cleavage between native islanders and residents "from away" in Grand Manan social relations (Marshall, 1999), a cleavage that significantly impacted all aspects of my research. While being "from away" gave me particular access that would have been denied an islander, as an outsider there were other issues of confidentiality and ethical behavior that demanded explicit attention. As Judith Stacey points out, "elements of fictionalization are intrinsic to ethnographic storytelling" (Stacey, 1990, p. 36). This fictionalized aspect in the documentation becomes part of our ethical conduct. The very nature of writing requires that restrictions be embedded in the process itself (Hopkins, 1993, p. 155).

The second borderland infusing my study can be described as *the absence or presence of a feminist perspective in understanding gender relations*. While my original commitment had been to a feminist ethnography, it quickly became apparent that patriarchy is neither unitary nor self-contained, and it cannot be separated from evolving social structures. "An understanding of gender inequality requires analyses both at the level of social structure and at the level of the individual" (Fox, 1988, p. 163). It was impossible to consider women's struggles in Grand Manan in isolation. Feminism became not only an epistemology, but a strategy, a pragmatic tool that was useful in some situations and not in others (Frohlick, 1999). Such a position "involves leaving aside a narrowly focused analysis of 'women' to ask questions about the 'other' side of gender" (Frohlick, 1999, p. 299). For the researcher, the sense of a borderland involving feminism, and concomitantly gender relations, applies not only to defining terms of reference for study, but also to the activities of everyday living.

One of the most striking dimensions of gender relations in Grand Manan is the spatial context through which public and private spaces of interaction reflect and control women's lives. For the outsider, there may be ambiguity about "acceptable" spatial patterns of mobility and interaction, especially as they relate to gender. Feminist ethnography forces the researcher to confront questions about local normative practices. "The social construction of gender difference establishes some spaces as women's and others as men's" (Blunt and Rose, 1994, p. 3), which, in the separation of public and private spheres of action, is central to the feminist struggle (Pateman, 1989, p. 118). Judith Stacey describes the importance of gender for her research, pointing out that "Being a woman inhibited my access to, and likely my empathy with, the complete range of male family experience" (Stacey, 1996, p. 26). It is an issue that I have been grappling with throughout my research on the island, and no doubt will continue to do so.

Tensions associated with the historical context of patriarchy and the space between *research and action* constituted the third borderland I inhabited within my research. The debates around feminism and the need to move forward within the emancipatory project through which women can be empowered have influenced my approach to research (see Gottfried, 1996; hooks, 1984; Jaggar, 1983). Women interviewing women as a research strategy is potentially activist because of the inherent role in consciousness-raising (Laws, 1986), bearing in mind that any attempt to speak for someone else is fraught with dangers of presumption as well as ethical and moral responsibilities of interpretations that accurately reflect the voices of others. Indeed, Robinson (1994) has argued that the authoritative voice of the researcher must be muted, though not eradicated, in the objective of eliciting meanings and interpretations of women's lives. The dilemmas implied in a commitment to action intensify the burden of "difference" rooted in lack of experiential understanding and shared meanings. For a feminist ethnographer, a focus not only *on* but *on behalf of* women carries significant political dimensions that molds the entire research process. As researchers, we are challenged by the intensely personal nature of ethnographic research that "places research subjects at grave risk of manipulation" (Stacey, 1996, p. 90).

Briefly – The Setting

Since Grand Manan has a fragile environment it is important to minimize the conflict of uses. (Grand Manan Rural Plan, January, 1999, p. 5)

An island community of 2,600 people, Grand Manan is in the Bay of Fundy near the border between the state of Maine and the province of New Brunswick, 90 minutes by ferry from the mainland. Dependent upon a rich and diverse fishery that includes herring, groundfish, lobster, scallops, sea urchins, and several smaller niches, the people of Grand Manan have a strong sense of identity rooted in family lineage and their shared experience of the sea. For two hundred years the women have looked after family and community while husbands were at sea for weeks and even months at a time. New technologies, especially related to aquaculture, changing markets, degraded fish stocks, and government regulations have altered the seasonal and daily rhythms of work activities for the men, with profound effects on the women.

While herring continues to be canned as sardines in the automated fish plants and more recently used as feed for the new aquaculture industry, since 1998 it has ceased to be smoked in the sheds that used to dominate the island landscape. The loss of the smoked herring industry has meant that the women have lost their traditional and most significant meeting places around the fish tables. Unlike the fish factory where a time clock restricts the flexibility of hours so important to young mothers, and where a noisy assembly line limits social conversation, until 1997 boning sheds used to offer opportunities for sharing family tragedies, day-to-day victories, and the many struggles that define life in a fishing community. While the herring weirs continue as significant locations of community and social networks for the men, the women no longer have their smoke sheds to generate and nurture their interactive webs of significance. As I talk to the women who will say, "But, we're nothing!", I see on the one hand an heroic strength borne out of generations of survivalist strategies, and on the other hand a disturbing sense of feeling completely powerless.

Experiencing Borderlands – Some Dilemmas in Doing Feminist Ethnography

Like an unintentional pregnancy, the fieldwork on which the book is based seemed to happen to me and to determine its own path of growth. (Stacey, 1990, p. 27)

Entering the community and soliciting contacts

Entering the community was difficult. After an afternoon of wandering around the wharves, buying a few groceries, and looking through pamphlets at the business center, several impressions begin to stir up uneasy feelings of self-doubt. Where were the women? How can I make contact? Even the language was strange. New words, rhythms, accents and syntax on Grand Manan, combined with uneasiness in my limited knowledge of the fishery, made me distinctly uncomfortable. I had difficulty even formulating a sensible question because I did not have the vocabulary to describe the "hook" on the "weir" to which the "drop" was attached. I soon learned to become self-mocking about my ignorance and ineptitude around the fishery. Making jokes about my inabilities seemed to smooth over the bumpiness of engaging in conversations on the wharves. For the most part, this was a successful strategy that acknowledged both my outsider status and my willingness to cede any perceived outsider "expertise" to their male-centered knowledge. The men were always happy to answer questions that incorporated both genuine curiosity and respect with a sense of their male superiority. The borderlands of insider–outsider, feminist perspective, and research versus action would all come to the fore in situations that implicitly questioned the naturalness of the patriarchal structures.

The ferry trip itself, I have discovered, is inevitably an invaluable entrée through casual meetings in the lounge or on the deck. On one trip I sat with a young couple who, under normal circumstances I might never have had a specific reason to interview. For an hour I listened as he described how he made his career choice as a fisheries officer, and how in his spare time he scavenges for discarded metal objects that he separates and sells to various recyclers. While I had met them briefly before, and they knew about my research, the opportunity for an informal exchange when I shared my own interests and heard about theirs allowed me to establish an unsolicited relationship that might be helpful in the future. In a small community, being known is not necessarily the problem; being accepted is. Such contacts help to build a foundation of goodwill and to create the possibility of future sources of information and understanding.

Even informal conversations may not be adequate groundwork for the more formal interview, especially in the early stages of the research. During my first year, someone told me about a fisherman who was converting his lobster boat, *Spiritwind*, for the summer season to take out tourists for "deep sea fishing." Wandering about the wharves, I noticed someone working on the *Spiritwind*. We started to chat, and our casual conversation led to an invitation to accompany him and his son the next day when they went out to pull their traps. It was a wonderful day, full of stories and information about the island and its people. Several days later I phoned his home, spoke to his wife, and asked if I might come over for an "interview" about changes on the island. She greeted me at the door, and led me into the living room, where I encountered a stiff backed, neatly combed man sitting, looking extremely uncomfortable. No jokes that evening, no long rambling stories; this was to be an "interview"; both the fisherman and his wife, a teacher, were on edge. By the time I left a couple of hours later the atmosphere was certainly more relaxed, but I was not happy with the information I had been able to collect. Although phone calls to arrange formal interviews are a useful introduction for some people, my experience on Grand Manan has been that "dropping in," or asking to drop-in when I happen to see them at church, for example, is far more likely to elicit a rich source of information.

Meeting women

The fun or, depending upon your viewpoint, the frustration of ethnographic research is that it follows few rules. As Stacey (1990, p. 27) points out "serendipity in the process is a more common and valuable research method than is often acknowledged." Arriving on the island, I soon realized that finding women to talk to would be more difficult than I had anticipated. The casual meeting places for men were off-limits to women. For the most part, the older women are at home or in the grocery stores or churches. The younger women are either earning money outside the home, or looking after children, only occasionally taking walks with baby carriages. That there are no sidewalks, no central public parks or commons for informal gatherings, and no leisure drop-in facilities for women has significant implications for social relations on the island.

One strategy that proved effective with the younger women, which seemed to encourage their willingness to meet with me, was to have group discussions that were already part of their weekly schedule (Davies, 1999). "Ladies Night Out" or Bible study groups served as venues for my inquiries. However, I soon learned that the apparent "efficiency" of these gatherings held particular hazards. In one case, the group was evenly divided between native islanders and those "from away" who had married into the island. Questions about changes at the school, about the changing nature of jobs available to women, and problems of youth, all elicited different sets of responses from each of the two groups. More importantly, in later conversations I discovered that much was left unsaid because of different perceptions, values, and concerns about island life that contained implied negative criticism that would have further alienated the "from aways" who struggle to be accepted. An "open" discussion it was not! For non-native islanders, there is an ever-present borderland of daily living that is constantly being negotiated and represents a fragile reality of acceptance.

This borderland between insiders and outsiders on the island was further illuminated in another group meeting that provided a contrasting interview experience for me. In this very fruitful evening, four women in their early forties, all university-educated with teenage children, had arrived on the island twenty years earlier as new spouses of native islanders. Their lively discussion over a period of three hours, revealed the depths of the insider–outsider cleavage in Grand Manan society, and, in particular, the difficulties faced by women who marry into the island culture. Group interviews are different from individual meetings in that the number of variables that can influence the nature and value of the information multiplies by the number of women in the group! At the same time, it is important to take into account the characteristics of the group that may or may not reflect existing cleavages in the community.

Yet another group showed me the danger of unrecognized assumptions. When I arranged to meet with a group of six native islanders, I assumed they would reflect some common understandings and a level of trust that would be helpful to me. I had not considered the, perhaps obvious, possibility that these "common" understandings can be asserted as a defensive wall, protecting island histories against the outsider. During "Ladies Night Out," we discussed issues such as women's roles in the home, their experiences in new tourism opportunities, and changing expectations of teenagers. The hostess for one of the evenings, who appeared to be lively, energetic, and even representative of the "strong women" of Grand Manan, seemed relaxed and engaged with all the questions I raised. It was only weeks later that I began to hear stories about the abuse she has endured for many years at the hands of her husband. Unlike other conversations and interviews I have had with individual women, during the entire evening there was no intimation of spousal abuse, despite the apparently widespread public knowledge of her situation. This example highlights the difficulties for the researcher of absorbing and adapting to the pervasiveness of the insider–outsider cleavage.

Three different group situations; three different responses; each with a different mix of people and perceptions focused on the intentions of the outsider–researcher. While the group of women "from away" had less invested in protecting island culture, and indeed seemed almost delighted that their stories might be heard, for women caught up in systemic abuse the group became a collective defense against the curious intruder. It was, in fact, the discovery of the underside of the "rural idyll" that was the darkest moment for me. I felt that I had "lost my innocence"! A comment by Stacey (1990, p. 272) echoes this feeling and more, when she describes her "gradual loss of naïveté about the ethical character of ethnographic research methods, which I discovered to be far less benign or feminist than I had anticipated."

Gendered responses

The borderlands of research activity in relation to activism were never more apparent than when I was confronted by the temptation to go beyond my "fieldwork as resistance" (Nast, 1994, p. 60), and to become a facilitator, and even an initiator, of resistance. During the third year of research on the island, I began to explore with women the notion of creating their own public space for casual, informal networking. In accepting the "artificiality of the distinctions drawn between research and politics" (Katz, 1994, p. 67), and feeling increasingly comfortable with the degree of acceptance I seemed to be accorded, I failed to understand the implications of my work for the researched community. Several stories illuminate the borderland that existed for me as a tension between non-participatory "neutral" recording, and the sense of urgency for women's empowerment.

During an interview with a young woman, "Susan," a school teacher "from away" who had moved to the island a decade earlier and married an island-born carpenter, Susan described her many concerns about raising their children and about the problems of student discipline at the school. While we talked, she was feeding a youngster in a high chair in the kitchen, and her husband sat listening in the adjacent living room. I asked her about the opportunities for women to get together informally, and wondered what she thought about the merits of some sort of indoor facility for women. Her immediate response was: "They wouldn't have time. They're tied to the home." But, just as telling was her husband's intervention from the next room: "Why would they want that?!" To this, Susan shot back, "You can drive up the island anytime and stop at the garage for coffee. Men don't have a commitment to stay at home."

Then she continued, thinking out loud, as she responded to my query: "There might be a need. But I don't know how much women would use it. I looked for a group when Philip was born. I tried to get a support group going for new moms. But there wasn't much interest. Annette [the social worker] tried, but it was in her office, at night. It couldn't work, especially at night when moms are tired. To see other women we have to call and make an appointment!" She remarked on a resource center she knew of on the mainland that seemed to be successful, and wondered if islanders might learn from their experience.

David (still in the next room): "Women always want to organize things. The first thing they would do is schedule who would make coffee and cookies each day. We just take our chances, and drop-in."

Speaking between rooms, I asked when was the last time that a male friend had dropped in at his house. David: "When Susan was away!" They both laughed, knowing the truth of it even as they seemed somewhat mollified to have to admit it. As the conversation continued, it was apparent that David saw many obstacles to the success of any initiative for a women's center. For Susan, the idea was enticing, but even as a woman "from away" who might consider challenging the patriarchal constraints of island culture, she noted that any such center would have to be "off the road" so that "parked cars couldn't be seen." "The women wouldn't want people to know they were there."

This conversation between Susan, David, and myself alludes to the unspoken resistances to women's mobility, but, more significantly, my own inappropriate and unacknowledged interference as an outsider. Furthermore, I realized that there was also an issue of unintended consequences. Not only did I begin to question the ethics of initiating and facilitating a women's center, there was also a concern about my ability to continue the research in the face of open hostility from significant sectors of the community. A mainland social worker cautioned me about the dangers for women who step out, in terms of abuse and violence at home. As the identities of men engaged in the traditional fishery are being threatened by declining fish stocks and disappearing markets, their insecurities play out at home. If at the same time women are seen to be gaining freedom from their traditional ties to the family, the situation may be exacerbated, with women as the scapegoats – physically, emotionally, financially, and psychologically.

While I have expressed my commitment to social action within the community in other ways, the situation around women threatens a fundamental normative structure. The idea of a women's meeting place would challenge a prevailing cultural norm that creates distinctive public and private spheres of men and women on the island. I drew back, and have not again broached the topic, except as it may come up during conversations about women's spaces and leisure activities. My deepening commitment to the well-being of the community that I see as directly related to enhancing and expanding the life experience of women, makes the borderland between research and action extremely difficult terrain to negotiate. Despite Porter's (1995) strong arguments for positions of advocacy, researchers must consider their ethical and moral responsibilities with respect to unintended consequences and any presumption of a "right" to interfere.

Gendered resistance

During many visits to the store for groceries, I would chat with a lively young cashier. I knew a bit about her background, and one day asked for an interview. We arranged to meet the next day, mid-afternoon, when she would be "doing the books" in the back office. The arrangement was important, both for her feeling of privacy and for maintaining a sense of informality about the interview. The impromptu nature of our meeting lent a "normalcy" to the meeting that seemed to enhance the comfort level for us both. That my vehicle would not be parked in her driveway, possibly provoking questions from curious islanders, I also considered to be an advantage, especially because she was known as being particularly out-spoken. Similarly, for the purposes of this narrative, the real identity of Janet is carefully disguised by inconsequential details. Her strength of character and reflective understanding of island values and structures provided me with a rich and informative interview. Her unequivocal, articulate comments about gender relations, and her stories illustrating the problems for women, provided insightful and disturbing views of a society caught in a "time warp." Janet watched as I wrote, talking not warily but with an awareness of both meanings and implications.

Describing an episode two years earlier, at a time when she and her sister, Laurie, were both involved in separation proceedings, she asked that she not be quoted by name. Her story was one of resistance, of openly challenging island patriarchal norms. Laurie had wanted to participate as an "extra" in the filming of a Hollywood movie that summer, but her husband would not allow it. Arguments ensued. His rationale was that the children needed to be looked after, the family should come first, and he could not do it since he would be off on the herring seiner. Despite his "orders," Janet and her mother offered to take over the babysitting so that she would be able to join the film crew on the movie set for each night for several weeks, which was what happened.

At the end of the shoot, a cast photo was taken, but Laurie was not allowed to have the photo in the house. Her mother had it safely stored until the separation was finalized months later. Story upon story during the two-hour interview revealed the lives of the two sisters and their mother, all now divorced. Three exceptional women, they are defying island norms and values, not only as defined by men but also as agreed to by women. For as Janet acknowledged, patriarchy is not merely men's oppression, but also involves women's acquiescence and complicity (Ortner, 1995).

Some Reflections

The subjectivities of both researcher and researched, then, are strongly implicated in the constructions and representations produced in texts. (Robinson, 1994, p. 217)

Negotiating the borderlands of feminist ethnography is an ongoing challenge both personally and for research. In various places throughout the preceding discussion I have referred to issues around gaining access to women's stories, confidentiality, ethical behaviour, and activism. In producing texts, we are faced with problems of "fictionalization" (Stacey, 1990), and "restrictions" (Hopkins, 1993), that challenge our own integrity and moral sensitivity. With Robinson (1994, p. 218), I agree that we need to speak *with* our subjects rather than *to* or *for* them, in a way that moves us forward "in the context of seeking modes of research and representation that disrupt – but do not suspend – the effects of positionality, the 'outside' perspectives." But her challenge is even more rigorous, when she calls for the mediations and interpretations to be inscribed as the object of investigation, rather than as the process in the investigation (Robinson, 1994, p. 220). There needs to be a displacement of the roles of researcher and researched so that the privileged voice of the questioner is not eradicated, but denied any space a priori.

In an early presentation of my findings, I showed my text to a long-time resident "from away," who, while acknowledging the validity of my emerging insights, also warned that islanders would not take kindly to a reinterpretation of their "white-washed history." As much as she had come to love her new home, she was well aware of the deceptions integral to the island myth of the rural idyll (Little and Austin, 1996). Unambiguously, she said to me: "Be careful. I don't want to see you vilified." Even as this constant tension defines every stage of the process, "we cannot launder everything" (Hopkins, 1993, p. 126). I have come to believe that there are no easy solutions.

For the feminist ethnographer who must negotiate the borderlands between the rural idyll and its darker underside, between the mythic significance of family lineage and growing reality of women's resistance, and between the need to empower and the ethics of confidentiality, there will continue to be tension. In becoming involved in feminist ethnography, we are inevitably engaged with the being and becoming of people's lives. In that, our moral responsibility represents a complex negotiation that contributes to the constitution and production of those lives. Similarly, our critical engagement with the "full texture of embeddedness" that defines the conditions and power of social relations "is crucial if researchers are to move . . . into 'the political' to effect change" (Moss, 1995b, p. 447).

It is not only men who live "in a time warp from the fifties" on Grand Manan. Our feminist ethnography cannot easily separate or disentangle the lives of women and men, the spaces of public and private, nor the activities of production and reproduction. Our research is always a borderland that may be guided by politics, but which must be infused with sensitivity and underlain by a strong code of ethics. Strong women deserve strong respect. It is our privilege and their generosity that allows us to be there.

RESEARCH TIP

At the Interview

- Ensure that both you and your subjects are positioned comfortably and close to your tape recorder. If you are not recording the interview, be sure to have enough space to take notes.
- Reiterate the voluntary nature of participation.
- Make sure that your tape recorder is working properly.
- Demonstrate flexibility in interview timing, taking into account the potential for interruptions.
- Acknowledge the need for breaks.
- Listen actively, e.g. smile and nod in response to your interviewee's comments rather than merely uttering "uh huh".
- Be patient, being careful not to force the pace of the interview – silences can be productive spaces.
- Be conscious of non-verbal cues and communication.
- Be respectful of the interviewee and their space.
- Promise only things that you can deliver.
- Express appreciation for the participant's time and energy.
- If appropriate, reiterate your openness to continued communication.

11

Negotiating Positionings: Exchanging Life Stories in Research Interviews

Deirdre McKay

To do research, you need to think about positioning yourself in the process of collecting data and writing up. In interviews, people will ask you to explain who you are and why your research questions are important to you. To write, you need to reflect on the explanations you give and the way you develop your questions. Each and every part of the research process is open to scrutiny, not just by others, but by yourself, too. I open up my research here by considering the dynamics of personal exchange between researcher and respondents in interviews. My strategies for working through problems of exchanging personal information, respecting friendships, and advancing the goals of feminist projects are a starting point for my analysis and critique, and yours, too. As you read my reflections on interview dynamics, you can think about strategies to position yourself within your own project.

Creating Common Ground

I spent 1996–7 in the northern Philippines, doing doctoral research on gender and economic development. My project examined the experiences of women who had traveled overseas as contract workers and then returned to the Philippines. These female OCWs (overseas contract workers) had often left professional jobs, partners and children in order to work as domestics or homecare providers in places as diverse as Hong Kong, Canada, Singapore, Italy, Saudi Arabia and the United States. While abroad, they sent money to their families. They returned with gifts and capital to invest in small businesses. As OCWs, they also brought home a set of experiences, memories and interpretative skills that they often cannot express within their communities.

Being a young woman, traveling on my own, I had something in common with them. Or, that's what several OCWs told me, saying: "I've traveled alone, just like you." Shared experiences of travel were a way of creating commonality in difference, a strategy for creating a space between us – a contact zone – where we could open up a dialogue. Such conversations happened in settings that framed them as "social interaction": public transit, storefront counters, over cups of coffee in people's kitchens, walking up the path from the bus stop. I always named myself as a researcher, but was often re-identified by my companion as, more importantly, an independent female traveler – like her. These initial discussions of femininity and travel began as gestures of friendship but I thought I needed to formalize them as "research."

Frustration with Methods

Using information from these social interactions, I decided to make guideline questions for a semi-structured interview. I asked two of my OCW friends to help me formulate my questions by doing an open-ended interview, recorded on tape. By transcribing their stories and comments, I would be able to identify the themes that emerged and terms they used to discuss them. The three of us set out to retrieve some of our previous conversations as "data" for my project. Then I listened to the tape and wrote up a list of questions.

As a strategy to design an interview format, this proved wildly unsuccessful! Each time I used my questions to frame a woman's narrative, I either shut down the conversation or elicited simple answers, with little interpretation attached. Off the record, open-ended discussions with returned female OCWs were much richer. We spoke about boyfriends, friendships, loneliness and the many frustrations of being a foreigner. Our exchanges had uncomfortably more of myself in them, but equally reflective detail and revealing comments from respondents.

Discouraged, I returned to the original interview to find the transcript filled with useful insights and quotable phrases. I saw that the successful interview worked as an exchange of autobiographies between my OCW friends and me, instead of probing their lives with a predetermined series of questions. At those moments where I gave them some details of my own story I found that respondents were most reflective and critical of their own experiences. When my gestures towards sharing their vulnerability disappeared behind a structured list of queries, the data vanished alongside the reciprocity. What I initially understood as an informal way of collecting background material and establishing rapport had become a key element of "the research."

The Realities of Research

I tried to contextualize particular exchanges as "research interview" and others as "just talk" but my respondents resisted attempts to create this distinction. They used all their interactions with me to map similarities and differences between our positionings and experiences. The space of dialogue between us never vanished, but changed shape, being constantly recreated from both sides. As friends and respondents, these women taught me that my entire field experience was my project, not just the activities I undertook with lists of questions. What I learned is limited by the level of comfort they found with me. The selection from the interview I offer below suggests how this relationship was negotiated. This excerpt shows how exchanging personal stories allowed resistance to pass through my own privileged outsider/researcher positioning. Highlighting this resistance situates the anti-colonial and feminist politics of research in personal relations.

Exchanging Migrant Stories

Ruth and Marilyn (not their real names) grew up together. Both worked overseas in their twenties and early thirties and they chose to be interviewed together. We did the interview at my house one evening over dinner, sitting around the tape recorder. From this conversation, I've chosen excerpts describing Ruth's experiences in Italy and her eventual return home. Her story begins with her first job as a housemaid near Rome.

> R: Her son would bring me along into class . . . because I could speak English and that was . . . different – show and tell, you know. And they would teach me Italian, but after two months I was getting very frustrated. I'm going. I don't know where I'm going, but I'm going! My friend Ida called me from Rome and I said, how 'bout my off day! They were nice people, protective and stuff like that, but I was just feeling that I was in jail. . . . So I didn't go back for a week . . . and then I realized "what would I be doing for a job?" So I stayed with Filipinos, women and men, in a *pension* – and boy, was there a lot of misdirected stuff, stealing, sex [*laughs*] going on in there. "Oh my gosh, this is worse than what I left, I'm going back!" I called my employer: "I'm coming back!!" And after a few weeks, confusion again. My pride was lame – you don't have your pride anymore, you know, you're working as a helper and then all this other stuff . . . and feeling so deprived of everything, you know? And I left again and went back to Rome . . . I was really desperate for companionship and, umm, my friends and I were just on the loose. Whoever speaks English, here we are – we want to speak English with you guys. We met these African guys. They speak English, right? "Wow,

they really speak English, let's go for it!" And then we ended up kind of staying with them in hotels and stuff like that . . . Another whole mess, you know. It's another whole mess. Gosh, it was devastating . . .

M: Was it for your rent?

R: Yeah . . . Well, this guy had a salary [*trails off*] And so, what else do you want to know?

D: Wow, that sounds really familiar, like my own time – when I was out in Africa. I went there with my boyfriend and then I split up with him . . . I was working, as a barmaid in this hotel bar and some of the girls who were working in the bar and we went out for drinks . . . One of the guys asked me out the next day . . . And then, no money, only tips and it was kind of hard to get away from him . . .

(CUT: *Deirdre and Marilyn discuss details of relationships while in hospitality sector jobs.*)

R: It's when you're like, illegal, you know? When you're desperate, you do the worst you can ever do . . .

M: Yeah.

D: Yeah.

R: At least I had a hotel – these guys were training in the airport in Rome . . . But, for me, later on, I felt like I was, used, you know? Finally we found out how you could go to an agency and find employers for yourself. You don't need anybody there to find work for you. Because we learned the language . . . And we also found jobs through other Filipina girls and the employers – they had friends who were looking for girls and so it went, like that . . . And I was able to get rid of this guy. I was like, never mind, I also speak Italian, I speak English . . . These guys were married. It was the lowest kind of life I've ever experienced . . . But I tried several employers . . . I stayed depending how I was treated . . . if I was paid on time, like that. And I finally found this family and stayed a year with them. I worked so hard with them. They gave me all the benefits and stuff like that. They wanted me. So they processed all my documents and stuff. So, you know, social security and health insurance . . . they processed it and they were being honest about it. The group I was with, I was the first to have my legal papers worked out!

    After a year, I met my husband and got pregnant, so I left, but I had to make sure there was another Filipina to take my place, you know?

(CUT: *Details of Ruth's marriage.*)

R: I never told any problems – it was so shameful. They're so proud of their little girl, she's out there, abroad, you know? She's gonna send a lot of money. We're going to be rich. Because that's what anybody who goes out of the country does – they're supposed to be sending money home. You think that you can manage it, you're from the center of town, not the far barrios, you speak English, you're educated . . . you know? But for me, that's not the case, the choices that I made – I was not able to send the money that they were expecting. And that's shameful, you know, not to come home then, and so I didn't have the money to come home and they died.

D: That's a really common thing for people who go abroad from here, isn't it? Not being able to talk about how hard the circumstances . . .

M: Yeah, especially during that time, the 1980s, when I was in Singapore . . .

R: I realize there is freedom in the relationship with God. How dare you judge me! You know where you stand and who you are. That's enough security to keep you going on with life. Now, my heart is heavy. At least, now, I am able to talk to people going outside of the country, working abroad. At least they have to get a glimpse, you know, of what they're going to find out there. What people expect. What employers expect and stuff like that, you know? Because we came out there and we weren't instructed on what we were going to do, they threw us into a totally different environment. And I was like: What's this? I'm going to have to do it on my own, nobody told me. It was a total shock. That's my goal: to be able to help. To help people prepare and inform them on what they're going to expect, going out of the country.

D: What would you tell them?

R: Well, it's always agency, agency to agency like that, when you can do it on you're own. Well, if you speak English, like most Filipinos, English is spoken worldwide and you can do it, you don't have to throw a large amount of money to agencies. I'm really, really mad because of my experience with agencies. If I could get one lady out of the devastation that's going on out there, running from agency to agency, I would feel better. What's the worst is the Filipinos ripping off Filipinos. Ummm. You know, it's just ridiculous. We all need money but you don't have to live off your fellow Filipino and lead a luxurious life . . . When you know the law, you have to be educated with the law, you can do anything if you have it in your hands. A lot of the employers don't do what they should for the helpers. It's the same all over the world . . .
    But it's hard, especially for Filipinos to speak out. We've been so suppressed through our culture. Mabain [shy, ashamed, embarrassed – DM] – who's going to speak out? You know? We cannot speak up for our own rights because of the way we grew up . . .

D: I wouldn't fault Filipino culture too much . . . You're isolated, you're in a different place and you don't go out, you don't know what your rights are. I think it's anybody – you need to have the contact, the support, the education . . .

M: But being mabain makes it worse . . .

R: Yeah, but then why cause problems . . . It's the colonial thing for us

Filipinos – if they're white, they must be right. That's how we were taught. That's how we grew up here, you know!

D: That's true in Vancouver, too, there's a racial thing. One thing the research [*gesture towards a draft of Pratt, 1997, on floor*] found is that the expectations of people at home weigh much more heavily on the Filipinas and the employers know this. The women who come from Europe, from Britain, they're seen more as tourists – having a working holiday.

M: Because they're white!!

D: Yes, and because they're there for the experience, to improve their English, not to send home one-third of their salary. So they're not as vulnerable.

R: They've got their money, their families, the girls are just earning for themselves . . . They don't get a phone call saying that your sister needs an operation, send money right away . . .

D: If you look back on it, would you still go abroad?

R: I would still go to Rome . . . But I wouldn't do some of those things again. Going there, it's a very lonely place, a very vulnerable place to be, in a strange country, but it made me strong . . .

Analyzing our Exchange

Reading this transcript makes me uneasy. I'm uncomfortable with the power relations operating in the space between us and concerned about the implications of re-telling these stories. Writing up the research seems to be the key act of exclusion that recreates the visitor/friends divide into researcher/subjects. In writing this piece, there were some tales I just did not wish to re-tell. So, like all "truths," the transcript is fictive – partial and formulated for this context.

Though many women from the Northern Philippines have worked in Singapore and Italy, personal histories reveal particular details. Ruth and Marilyn asked me not to use parts of their stories for my research. The circumstances of Ruth's marriage and Marilyn's hospitality work would identify them to their community and our mutual friends. Having made these cuts, I'm left with a dilemma. Should I present my narrative of working on a tourist visa in a Johannesburg bar as a stand-alone confessional? In the end, I decided that alluding to it, without giving all the details in the transcript, would be enough to outline the dynamic I'm discussing here.

Writing up your research experience involves drawing a line separating "the story" from the rest. I find it difficult to choose where to draw this line. I want to balance the confidentiality of my respondents with my desire to present the politics of the research interview. I made my decisions on the basis that I wanted to focus on the way that the research interactions might be analyzed, not the "truthfulness" or transparency of the narrative.

Both differences and commonalities in our stories emerge in the transcript. As migrant women, we were commodified as aesthetic bodies, worked illegally in the service sector, and struggled to maintain ties with our sending areas. Like me, Ruth negotiated part of her journey through relationship with a "boyfriend." The relationship provided economic support and companionship, but left her a bad feeling. She had objectified and exploited her own femininity. But she had also moved from being the English-speaking curiosity in a child's show-and-tell to a place of self-respect and strength. Her experience gives her critical insights into her own culture, the ongoing experiences of colonization and the lived meanings of gender, race and ethnicity for Filipino women entering a global labor market. For Ruth, participating in my project was a chance to speak plainly, to some extent, in a context where the information might have an impact on someone else. Our interview was also an opportunity for her to speak back to colonization. We experienced differences in perceptions of our ethnicity in our host society, racial privilege in particular contexts, and linguistic ability. Ruth's comments highlight how these differences reflect uneven and hierarchical relations of ethnicity and race.

How could my analysis attend to both differences and common ground, without one obscuring the other? To theorize autobiographical exchange, I read about personal narratives in feminist theories of interview practice and feminist geographical methods.

Autobiographical Exchange in Feminist Theory

Offering stories of your own experiences as a way of eliciting the same from others is considered a feminine strategy. Participants in such exchanges can identify overarching themes, as well as personal or group-identified differences. This kind of talk can be therapeutic. By revealing previously hidden experiences, women can explore the impact of cultural categories, social processes and global economies in their lives. Sometimes these explorations lead to expressions of resistance. In other contexts, women examine the limits of their acquiescence to particular situations and attempt to create shared understandings of the compromises they've negotiated. Female respondents often bring expectations of therapeutic or political outcomes to research conversations (Anderson and Jack, 1991; Lather, 1986; and Opie, 1992). Feminist scholars have, for the past two decades, called for methodologies that anticipate female respondents' desires to find personal affirmation and social support within the research interview.

Likewise, researchers are asked, and feel obliged to answer, some very personal questions about their sexual and emotional biographies (Twyman et al., 1999). Feminist researchers cannot exclude this kind of talk from interactions with respondents. In conventional methodologies, exchange of biographical information occurs, but is classified as "establishing rapport" rather than data. The researcher's biography lies outside the research while the respondent's is made available for analysis. Oakley (1981), in what is now a classic paper, initiated the argument for interactive self-disclosure as part of a dialogic and collaborative feminist research process.

Stacey (1988) highlights the dangers inherent in self-disclosures made during ethnographic interviews. She argues that these egalitarian interview practices are problematic because they produce apparent truth about the "subjects" of research, without making the researcher's own story equally vulnerable to readers' interpretations. Feminist researchers are likely to emphasize empathy and alliance with their respondents and neglect to report women's resistance to the intrusion and exploitation of the research itself. Stacey is concerned that focusing on reciprocity creates a false version of solidarity and commonalities between women. By emphasizing transparent, mutual understanding, the researcher could obscure conflicting class, ethnic and other interests and the power relations of the research process itself.

Feminist geographers have begun to include autobiographical narratives in their research reports. Personal information on the author, in the terms shared with the respondents, enables the reader to examine the data collected and the roles played by researchers and respondents (see for example Twyman et al., 1999). Just as no gesture towards reflexivity will ever achieve complete transparency (Rose, 1997), no autobiographical narrative will reveal or address the full interplay of power relations in the research process.

However, identifying a researcher's autobiography as "data" produces an account of the research that contextualizes research more broadly. Yet where researchers offer autobiographical detail in their writing, but separate their stories from those of their respondents, the terms of the exchange remain outside the analysis.

To resolve these tensions, I decided to focus on resistance to my presence as researcher within the interview and to the politics of research itself. Resistance highlights respondents' negotiation of the relations of exploitation that Stacey is concerned that feminist ethnographies may conceal. By exploring resistance, I might be able to better describe both the reciprocity and inequalities between researcher and respondent. To do so, I decided to approach the exchanges of autobiographical information as autoethnography.

Autobiographical Exchange and Autoethnography

Autoethnography is "self-writing-culture" – writing of a self back into a culture that has been bounded and determined by the descriptions of previous authors. Former research subjects write themselves back in to the official ethnographic record, resisting and renegotiating their representations as produced and circulated through "research." Unlike the specialized academic system of description it critiques, resistance autoethnography may be understood by a much wider audience. Using the everyday strategy of telling personal stories to explain events and contexts makes autoethnography accessible.

Personal and autobiographical responses to ethnographic descriptions can also be described as autoethnographic. Personal autoethnographies critique cultural values that norm one's self-concept. These two forms of engagement, resistance and personal positioning, are easily intertwined in cross-cultural research. Analyzing a research interview as autoethnography could provide a way to acknowledge both resistance and acquiescence to the research process.

Autoethnography as Resistance

Pratt (1994, p. 28) uses the term autoethnography to mark the special characteristics of texts and performances in which a colonized people attempt to describe themselves in ways that engage with their representations by others: "[An] *autoethnographic text* ... [is] a text in which people undertake to describe themselves in ways that engage with the representations others have made of them. Thus, if ethnographic texts are those in which European metropolitan subjects represent to themselves their others (usually their subjugated others), autoethnographic texts are representations that the so-defined others construct *in response to* or in dialogue with those texts."

Autoethnography is a characteristic response of people within the contact zone – a space where colonized people respond, resist, collaborate, adapt, communicate and imitate in their attempts to engage the categories and idiom of the colonizers. A contact zone is the space of encounter – where colonized and colonizer are present together and to each other. Here, ethnographers conduct research on the colonized and categorize them. The contact zone is also the place where colonized peoples respond to research and the categories used to describe them. The space between researcher and respondents in a cross-cultural research interview is a good example of a contact zone.

In research interviews, autoethnography re-asserts the native self and engages a broader audience, beyond the researcher. You can hear this in Ruth and Marilyn's comments on colonialism (see pages 191–2). Against my assertion of commonalities among female migrants, Ruth and Marilyn insist on specific difficulties faced by migrants from a formerly colonial nation. Their critique of Filipino culture engages with the stereotype that Filipina migrants are "passive" and do not stand up for their rights. In their experience, English language skills make them marketable as migrants. However, their English came with an understanding of racial hierarchies instilled by a colonial educational system. They are not completely rejecting the representation of Filipina migrants as reluctant to speak out, but explaining this response in terms of colonial encounters.

Autoethnography as Personal Positioning

Particular self-reflexive autobiographical narratives are also autoethnographic. Reed-Danahay (1999, p. 9) defines autoethnography as: "a form of self-narrative that places the self within a social context. It is both a method and a text, as is ethnography. Autoethnography can be done by either an anthropologist who is doing 'home' or 'native' ethnography or by a non-anthropologist/ethnographer. It can also be done by an autobiographer who places his or her life story within a story of the social context in which it occurs."

Autoethnography thus marks a new type of autobiographical writing where the author's goal is not to discover hidden aspects of the self but to rewrite personal history in a way that (re)creates a collective identity. Such autobiographies use the performance of language to critique and perhaps subvert or expand the cultural categories that authors identify as structuring their experiences. Their authors position themselves, through the stories they tell of their lives, at the critical intersection of histories, cultures, institutions and texts. An autoethnographic voice "concentrates on telling a personal, evocative story to provoke other's stories and adds blood and tissue to the abstract bones of theoretical discourse" (Ellis, 1997, p. 1). Anticipating and intending to elicit other peoples' stories with their autoethnographies, these authors engage their audiences in a reworking of group identities. To work autoethnographically is thus to engage in a struggle to politicize one's personal history – to mark the personal as political. Writing this way requires the exploration of one's self-knowledge as a social construction. An example of this is Nicotera's (1999, p. 430) description of her academic autobiography: "this autoethnography presents descriptions of events in my life when I was consciously aware of my existence as simultaneous subject and object." A similar awareness emerges in Ruth's description of her participation in the show and tell exhibit in Italy (see first line of narrative). Though the family was nice, she felt like she was "in jail" and left the job. Her later comments (page 192) on her education – "if they're white, they're right" – explain the social context in which she understood her employer's request that she go to school with their son. Anticipating that other women will want to go abroad, Ruth wants to share her story as a way of opening a dialogue on Filipina identities, experiences overseas, and colonial enculturation (see page 191).

Ways You Might Use Autoethnography

Autoethnographic work is based on the double awareness of self as subject *and* object and self-knowledge as social construction. By engaging in self-reflection on identity as historical, cultural and political, anyone can write autoethnography. As a researcher concerned with positioning a "self" within the research project and process, writing this way could be a useful strategy. You could try it before, after and during fieldwork and see if and how your self-understanding changes. Keeping your autoethnographic exercises would give you resources to position yourself in your writing later on.

Autoethnography is also a useful tool for the analysis of research interviews, providing a balanced and nuanced description of the interview context. To suggest the potential of such an analysis, I provide autoethnographic readings of two moments in the transcription excerpt.

First, I ask you to consider the moment where Ruth closed down her narrative (p. 190, line 5). After her answer trailed off and ended in a question, I shared my own experience in Africa (p. 190, line 7), offering what I now understand to be autoethnography to sustain the conversation. My (unconscious) goal was to encourage her to feel comfortable with the similarity of my experience. Ruth responded by continuing with her story. Later, she spoke back, directly, to the racist overtones within my assertions of commonality. Ruth's expressions of resistance to my identification with her experience are a key part of reflexive research practice. In this interaction, my personal autoethnography elicited anti-colonial autoethnography from my respondents.

My gesture towards the research results on the floor marks the second autoethnographic moment. Both women had scanned text before the interview as I explained that it would appear in an academic journal. I spoke briefly about my contacts with Vancouver's Philippine Women Center and Gerry Pratt's co-operative research project on Filipina domestic workers in Canada. Because I situated my project within this field of feminist research and activism, Ruth and Marilyn understood our interaction to be an iteration of other conversations with women who had similar experiences of migration. Ruth directs her comments to what she imagines might be discussed in a broad field of research and representation – ideas about Filipinas and Philippine culture. In this moment of resistance autoethnography, Ruth corrects what she thinks might be misrepresentations of Filipina experiences. The political promise held within the text-as-research-product sustains the dialogue between researcher and respondent. That researchers can take personal stories and turn them towards political goals, as exemplified by Pratt's paper, suggests why, despite her ambivalence and resistance, Ruth would continue the interview.

Work In Progress

My analysis of the excerpt of autobiographical exchange is unfinished. As ethnographer, author and person, I'm ambivalent about this methodology that I caught myself using, unplanned, unprepared for. I can't claim to have worked out all the issues, or even to have identified all the salient ones. Writing about this excerpt as a work in progress addresses some of my uncertainties in autobiographical writing and my misgivings about the realities of my field research.

By outlining a situation that every feminist researcher faces at some time or another, I intended this piece to evoke emotions about exploitation and reciprocity in the research process. Presenting you with a partial analysis leaves room for your own stories of tensions between friendship and research, questions and answers, self and other. Will you, as I tried to do, take the risks of disclosure and exposure and call your exchanges of autobiography a methodology for creating autoethnographic commentary? Or, will you withdraw your personal exchanges from the "data" you collect and thus distance yourself from your subjects? Would you, too, feel ambivalent but relieved by deleting "research" versions of your personal history from published interview excerpts? There's no correct course of action I can suggest for you; I can only ask you to think about when and where you would make such decisions.

RESEARCH TIP

Post-interview De-briefing

- Regroup and eat something that pleases you (for energy).
- Label tapes, and duplicate tapes (if necessary).
- If interview was not recorded, recount in detail the entire session as much as possible.
- Make notes about the interview session, including your impressions on the rapport, the content of the interview, things you might have forgotten, insights that came to you during the session, how you are feeling (tired, enthused, excited, nervous), and so on.
- Transcribe the tape as soon as possible.
- If employing a transcriber, get the tape to the transcriber as soon as possible and agree on completion dates.
- Talk with a colleague about any uncertainties encountered in the session.
- Acknowledge the interview in some way, e.g. telephone call, thank you card.

12

Interviewing Elites: Cautionary Tales about Researching Women Managers in Canada's Banking Industry

Kim V. L. England

The amount of resources that it would take to answer your questions is tremendous. It would require a lot of digging and we're not going to do that.

Having read the types of questions and the level of detail you are asking for, we are not in a position to participate at this time. Although we have done a fair amount of work around the advancement of women at [name of the bank], the research and data is proprietary and not for external distribution. So, we realized that we wouldn't be able to share much with you, and didn't believe our participation would add much value.

These two quotes are from responses I received to requests for interviews with high-ranking managers in human resources at Canadian banks. In this chapter I look at the process of interviewing "corporate elites" based on self-conscious reflections on some of my research projects, especially my current work on the gendered geographies of Canada's banking industry. In recent years, geographers have shown a growing interest in the study of elites as elites, especially in economic geography (see, for example, the special issues in *Environment and Planning A*, Cochrane, 1998 and *Geoforum*, Herod, 1999). This is a departure from the more usual focus by geographers on the politics, ethics and practicalities of studying relatively less powerful, and even vulnerable groups. While some of the concerns are the same in both cases, interviewing elites does raise different sorts of issues than for researchers studying less powerful groups. In this chapter I offer cautionary tales and what I hope is constructive advice regarding fieldwork about elites.

Gendering Workplaces, Gendering Work Practices

In feminist geography, the field of paid employment has been a fertile area of research and one of its most enduring. In some of the more recent work in this field the workplace itself has come under closer scrutiny. Several feminist geographers identify organizations as gendered and workplaces as sites where identities, power and knowledge are discursively (re)produced. They argue that the everyday social and cultural practices of work are gendered and embodied (see for example Hanson and Pratt, 1995; McDowell, 1997a; Halford, Savage, and Witz, 1997). In my research about women managers in the Canadian banking industry I adopt the framework of gendered organizations and gendered and embodied work practices. Canada has a national system of banks and is dominated by what are known as the "Big Six" banks (Bank of Montreal, CIBC, National Bank, Royal Bank, Scotiabank, and the Toronto-Dominion Bank). In 1996, 93 percent of all bank employees in Canada worked for one of the "Big Six" banks. The "Big Six" are either headquartered in Toronto, or have a significant presence there. Thus, 35 percent of Canada's "Big Six" employees are based in Toronto, but they make up only 15 percent of Canada's total labor force. Banking is highly feminized (75 percent of workers are women), compared with the labor force as a whole (45 percent). Moreover, Toronto has a high concentration of women managers relative to men in banking and relative to women in other industries. I decided that all of this made Toronto a good place to explore whether, as Eleonore Kofman recently put it "major cities offer women more opportunities for social mobility, including movement from professional to managerial sectors of the service class" (1998, p. 290).

I knew, from an analysis of newspapers and news releases, that since the mid-1980s the "Big Six" banks had introduced various policies and programs to encourage the promotion of women managers, and that the numbers of women in managerial positions had increased over the same time period. The mid-1980s were also significant because in 1986, the Federal government introduced the Employment Equity Act covering federally regulated employers like the "Big Six" banks. Specifically, the 1986 Employment Equity Act covers crown corporations and federally regulated employers in the banking, transportation and communications sectors (with at least one hundred employees). The purpose of the Act is "to achieve equality in the workplace for women, Aboriginal peoples, persons with disabilities and members of visible minorities. In the fulfillment of that goal, employers are asked to correct disadvantages in employment experienced by the designated groups by giving effect to the principle that employment equity means not only that they must treat people the same way, but also that they must take special measures and accommodate difference" (HRDC, 1998, p. 1). In 1995, the Act was replaced by a new Employment Equity Act, which clarifies existing employer obligations to implement employment equity. (See http://info.load-otea.hrdc-drhc.gc.ca/ ~weeweb/lege.htm for more information about the Employment Equity Acts.)

At the outset of my research project on employment equity and Canadian banking I set myself three broad research goals. First, I wanted to document the increased presence of women in the managerial ranks the "Big Six" banks. Then I planned to explore how and why the "Big Six" introduced policies and programs to advance the promotion of women, and find out the extent to which the decisions to develop these policies were influenced by the introduction of the Employment Equity Act in 1986. Finally, I wanted to see whether the banks' policies and programs (that I presumed were developed in their headquarters, which in most cases are in Toronto) had been portable "as is" to other locations across Canada, or whether (and how) they had adapted them to different locations across Canada.

Mixed Methods, Triangulation, and "Shameless Eclecticism"

"Ethnographers have shown themselves to be willing to employ practically every technique available to the social scientist: sample surveys, informants, census, historical documents, direct participation, first-hand observations, descriptive linguistics, correlation techniques, psychological tests and so forth: 'shameless eclecticism' and 'methodological opportunism' are defining features of the ethnographer" (Jackson, 1985, p. 169).

I am sure I am not alone in feeling that these words still capture the experience of doing fieldwork. Certainly in many of my research projects I have been a "shameless eclectic" and "methodological opportunist," using a good number of the techniques that Jackson lists! Moreover, sociologists Rosanna Hertz and Jonathon Imber, in their important edited book *Studying Elites Using Qualitative Methods*, point out that the "best research on elites has utilized a combination of methodological approaches to deepen the research findings. . . . Rather than assuming that qualitative and quantitative research methods are always at odds, the multi-method approach casts constructive doubts on relying on the use of any single source of data or method" (1995, p. ix). And in their discussion of postmodernism and organizational research, Martin Kilduff and Ajay Mehra (1997, p. 458) argue "that no method grants privileged access to the truth and that all research approaches are embodied in cultural practices that postmodernists seek to make explicit." Of course, this echoes the current trend among feminist geographers towards flexibility about the correspondence between certain philosophical orientations and particular methods, rather than some sort of methodological orthodoxy (see for example G. Rose, 1993; England, 1994; Lawson, 1995; Moss, 1995a).

I find that a theoretically informed blending of qualitative and quantitative is the most effective way for me to approach empirical research. For example, one of my previous projects dealt with the role of local clerical labor markets in the location decisions of firms in Columbus, Ohio employing large numbers of clerical workers. I tackled my research using a variety of overlapping data sets and approaches. I "triangulated" census tract data analysis with questionnaires and in-depth interviews with both clerical workers and personnel managers (see England, 1993, 1995). At the time I was planning that project I was greatly influenced by critical realist methodology, in particular the concept of extensive and intensive research (Sayer, 1985; Sayer and Morgan, 1985). By uncovering general patterns and common properties among a representative sample of the population, extensive research aims to answer questions like "how many?" "how often?" Intensive research focuses on the question of "why?" and is concerned with how causal properties are manifest in a particular set of cases and provides explanatory power. The basis of my extensive research was the census data analysis and analysis of some other secondary data and I envisaged my intensive research as set of case studies of a number of firms involving interviewing the personnel managers and a number of their clerical workers. However, after entering the field, I soon realized I would need to really take seriously my identity as a "shameless eclectic" and a "methodological opportunist." I approached a handful of firms who agreed to let me interview their personnel manager, but refused to let me interview their clerical workers. In a couple of instances, that I wanted to interview their clerical workers led the firm to decline to be involved in my project at all. They usually cited reasons revolved around employee privacy. Of course, this might be the case, but I wondered whether it was because they did not like some of my questions (I showed them the list of questions I proposed to ask clerical workers) and were concerned that I might encourage unionization (only a couple of the twenty or so firms where I interviewed personnel managers were unionized). It became very clear I was going to have to make do with whatever I could get. So I began suggesting an alternative – a short questionnaire – to be distributed to clerical workers by the personnel manager and each respondent returning the questionnaire in a sealed envelope. This proved to be more satisfactory to firms. One personnel manager offered to give me a print-out of the firm's clerical workers by sex and the zip code of their home address (one of my research questions was whether the firms draw on highly localized clerical labor markets). I snapped up the offer. And a number of other firms were willing to provide me with similar data. But I still wanted to interview women who worked as clerical workers. In the end, I found respondents by combing through the *Polk City Directory* to find clerical workers who lived in the same suburbs in which the firms were located, and a few of those women happened to work for the firms where I also interviewed the personnel managers.

The experience of the Columbus project, among others, behind me, it was always my intention to enter the field in Toronto fully anticipating the revival of the "shameless eclectic" and "methodological opportunist" when it came to the women bank managers project. I initially planned to use secondary data and publicly available documents (for example annual reports from the "Big Six" banks and material from the Canadian Bankers Association) to trace the growing presence of women in the managerial occupations at the "Big Six" banks. I "triangulated" that information with material gleaned from newspapers, news releases and business print media. I then hoped that the banks would give me access to documents about their employment equity and diversity programs (e.g. reports to employees focusing on the banks' taskforces on the advancement of women). Finally, I wanted to conduct face-to-face interviews with human resources managers specializing in employment equity and workplace diversity at the "Big Six" banks. When I was planning this research, I had hoped to speak to a sample of women and men managers at the banks about their careers and balancing their work and home responsibilities, but my initial inquiries indicated that there would be great resistance to this, both by the banks and by the managers themselves. So I did not pursue this aspect of the research. At the time, I thought my research goals were quite modest and relatively uncontroversial. Was I wrong! Documenting the numbers of women in managerial ranks was more time-consuming than I expected, but relatively straightforward, and getting hold of internal documents proved to be far easier than I anticipated. Trying to get managers to agree to be interviewed was very, very difficult. For example, in one instance I had the following exchange (the manager is a woman):

Manager: Either in our previous conversation or in this letter here you said something about wanting personal opinions, but I am representing the bank.

Kim: Well, I meant that I'd be asking you a question like "If you had a piece of advice to offer women wishing to reach senior management what would it be?" or "If you could create your own employment equity program what might it look like." So I thought it better to describe that as personal opinion than as representing the bank. But I'm not going to ask something inappropriate like "Come on, tell me what you really think about the bank's policy!" [Laughs.]

Manager: Hmmm. [Long pause.] Yes, that would be inappropriate. [Short pause.] Questions about personal opinions would be inappropriate. And anyway, programs cannot and are not built on personal opinions.

Kim: Yes.

Manager: Well then, we're on the same wavelength.

This exchange took place on the telephone. It was my fourth conversation with this particular manager. Approximately four months before this conversation I telephoned the personnel departments of the "Big Six" to ask for materials about the banks' programs for employment equity. I spoke with the manager quoted above during this early stage of the research process. She was very pleasant and helpful and chatted with me for almost five minutes about the bank's policies and programs. Within a week I received a very large packet of materials from her. As I thought she seemed likely to be willing to be interviewed, I mailed her a letter describing my research and requesting an interview. I included a list of potential questions I might ask her. She then phoned me to ask me to send her the specific questions I wished to ask her. I did so. We then had a further telephone conversation discussing which questions she was not prepared to answer and how she wanted some of the other questions reworked. I complied. At the end of the telephone conversation in which the above exchange took place she told me that despite her initial interest, she would not participate. She told me that "the amount of resources that it would take to answer your questions is tremendous. It would require a lot of digging and we're not going to do that." I concluded that interviewing elites is certainly not for the faint-hearted. It can be very disappointing, not to say deflating. In fact, I had to develop what Joan Cassell in her study of American surgeons described as "brute persistence and blind compulsivity" and like her I kept "pushing, and trying, and hoping, and smiling, and pushing some more. For this, a researcher needs a thick skin and a certain imperviousness to rejection" (1988, pp. 94–5).

Are Elites Seen but (Willfully) not Heard?

In the past Canadian banks were notorious for being particularly inaccessible and very guarded. For instance, until relatively recently most did not even have websites. Colleagues warned me that I should expect difficulties gaining access. In addition, a rumor was circulating that in the past a graduate student had been given extensive access to one bank, and wrote a dissertation that portrayed the bank in a very negative light, and now the banks were unreceptive to academic researchers. (I no longer recall from whom I first heard this rumor. More recently I have been told that this "rumor" might have been circulating for decades and perhaps is worthy of promotion to urban myth status! Nevertheless it continues to work to deter all but the most determined and stubborn of researchers.) All of this served to make me anxious about my research because the harsh reality is that there is an extremely small number of people for me to approach for interviews – I am dealing with corporate elites employed at six banks. So rejection is more devastating to my research than if I were dealing with a more plentiful supply of potential interviewees. The rejection I described in the last section spurred me on to do as much as I could to secure interviews with the other banks. Fortunately (and unlike many other sorts of elites or even many other industries) there is now a wealth of secondary data in the public realm about the banks and about many of their senior managers for those willing to spend a great deal of time surfing the web and ferreting through libraries, archives, and the like. For example, the banks' websites usually include things like recent speeches by the CEO, and news releases. And commercial websites (such as www.hoovers.com), and the annual reports of the banks and other trade reports (for example, those published by the Canadian Bankers Association) contain a huge amount of information. In addition, I found that many bank executives offer important details about themselves for entries in volumes such as the *Canadian Who's Who* and *The Financial Post's Directory of Directors*. All of this is in addition to relatively frequent stories about the banks' employment practices in newspapers and business magazines.

However, this sort of carefully controlled visibility belies how difficult it is for a researcher to actually meet face-to-face a member of the banking industry elite. As with other elites, and as members of powerful companies with national and international reach, the managers are able (i.e. they have the power) to construct and maintain self-protective and formidable barriers to deter unwanted "outsiders" like me. Michael Useem notes that "the wealth and authority of corporate executives can be daunting terrains for the first-time visitor. Their status barriers may seem virtually impregnable, especially when contrasted with the thinly guarded and well-trodden paths of the poor and powerless" (1995, p. 20). Elites in general may insulate themselves from unwanted intrusions for many reasons. They may be defensive or protective because they do not wish to be studied or to have their "sensitive" decision-making processes revealed and scrutinized, or they may be uninterested in the project, or too busy and wish to guard against "unproductive" use of their time.

Moreover, as Robert Thomas remarks, "most businesses, no matter how small, have gatekeepers who keep an eye on the comings and goings of strangers. Large corporations, especially ones with trade secrets to hide, have gates, guards, and security devices. . . . You cannot just walk into an office suite and expect to strike up a conversation or hang out and observe the scene" (1995, p. 5). In all but one case where I was granted an interview by a "Big Six" bank manager, I arrived and a receptionist – arguably a "gatekeeper [keeping] an eye on the comings and goings of strangers" – telephoned the manager with whom I had an appointment and I was asked to wait in a reception area until the manager came to collect me. In the other case, I arrived at the office suite that contained the manager's office and was confronted not by a receptionist, but a huge frosted glass door and a telephone without a listing of telephone numbers. Scrambling through my bag, I congratulated myself on actually having brought the manager's business card and I phoned her number. She greeted me pleasantly and then proceeded to make remarks that made it clear she thought we were going to talk by telephone. Obviously I surprised her by physically showing up for the interview. She buzzed me through the glass door, and then several minutes later emerged from the labyrinth of offices, again expressing her surprise that I had shown up in person. I felt deeply embarrassed by this and worried that the interview would go badly. But, thankfully it went well. She told me "we get several requests for interviews every week, we don't agree to do them all." She then paused, smiled and raised her eyebrows, and continued, "mostly they are journalists and mostly they want to talk to me by phone." I took this to mean that one reason she agreed to talk to me was because I am not a journalist, whom, I am guessing, she views with a degree of skepticism.

My point is that gaining access to elites is hard work. It is about continual negotiation, bargaining and compromise. Hertz and Imber (1995) suggest gaining access to business elites is notoriously difficult for social scientists. However, many of the chapters in Hertz and Imber's book are written by academics in business schools or in management studies or business administration. These authors and pieces written by other similarly employed academics, do stress the difficulties of gaining access, but I cannot help wondering whether it was a good deal easier for them than it was for me as a geographer (I frequently had to field the "how is this geography" question). Many of the authors suggest that researchers draw on professional or personal contacts, or say that they intend to use their interview material for teaching or training future managers, or let organizations know that the researcher or the researcher's institution is a source of future employees and consultants. These potential advantages do not really apply to me. Indeed, in one bank I asked about circulating a questionnaire among their managers and offered to include questions that they might be interested in. I was told that they paid outside management consultants to do that kind of thing and also had a long-standing relationship with a professor from a prestigious Canadian business school who specializes in the advancement of women.

I worked incredibly hard to gain access. I used any means at my disposal to gain access. As I have no personal or professional links with the banks, I had to make a lot of "cold calls." After many years of doing qualitative research, I have much experience of being a "shameless eclectic" and "methodological opportunist." But my bank managers project meant that I also quickly became a "shameless opportunist." For example, I took advantage of my colleague, Gunter Gad's suggestion that I use his name to get my foot through what seemed like an impenetrable door and mercilessly pumped an archivist for information once he seemed to connect with me because of our shared "Englishness." Gaining access also takes a long time. It took me close to eight months to gain access to the people I interviewed in 1999. As I write this chapter, I am still involved in the process of negotiating access to another bank. In this case, I am yet again a "shameless opportunist" having accepted Gunter's offer to accompany him on an interview or have him ask questions on my behalf.

Many scholars claim that their success (or failure) depended to some degree on timing, luck, chance or serendipity (McDowell, 1998; Parry, 1998; Herod, 1999). This too has been my experience. For instance, one bank had just received a prestigious award for their progress in promoting women into senior management, and their personnel managers were very keen to be involved in my project. In the case I described earlier a manager seemed reasonably interested in my project and then (rather abruptly) ended what had been lengthy negotiations. A couple of weeks later, the banner headline in the business section of one newspaper provided a possible explanation – the bank had announced that it would be "shedding jobs." I like to think that this explains why she spurned me! However, I think it is very important not to put too much weight on luck and chance, or lack thereof. Negotiating and (hopefully) gaining access requires lots of careful preparation. It requires thinking through why you want the interview with this particular person and being able to explain clearly (and without jargon) why they should grant you the interview. I found it important to do this because some managers needed to be convinced that I really needed to interview them rather than their assistant or a more junior colleague, or in one instance, a representative from the public relations office. Negotiating access required that I convince the managers that I absolutely could not get the information elsewhere. In an effort to secure interviews I sent out letters addressed to the manager I wished to interview (I found it reasonably easy to obtain the name of the manager with a telephone call to the headquarters of the bank). My letters included a brief explanation of the nature of the research and emphasized the importance of the manager's personal input in order for the project to be successful. I customized letters to indicate that I had done my homework. For instance, I made sure I mentioned the names of the bank's specific programs, policies or documents dealing with the advancement of women. I tried to make it very clear that I was not going to waste time by asking about material that is available from other sources, and that I had customized my questions based what I could not obtain from other sources. I now firmly believe that it is simply not possible to over-prepare for an interview!

Ethics, Positionality and Power Relations in the Field

For some time now, feminist meditations about the research process have confronted the impartiality and objectivist neutrality of quasi-positivist empiricism. Along with several other feminist geographers, I reject the notion of researcher as the omnipotent expert in control of both the passive researched and the research process (England, 1994; Katz, 1994; Rose, 1997). The strict dichotomy between researcher and researched is also rejected, and there is a careful consideration of the consequences of the researcher's interactions with those they research. Researchers are a visible and integral part of the research encounter, and, as Liz Stanley and Sue Wise (1993, p. 157) put it, "researchers remain human beings complete with all the usual assembly of feelings, failings, and moods." And, of course, the same applies to the researched. Neither I nor the researched have fixed subject positions.

Several feminist geographers, myself included, write in favor of pursuing a supplicant relationship with the researched, and seeking reciprocal relationships based on empathy, mutuality and respect. The appeal of supplication lies in its potential for the researcher to cope with asymmetrical and potentially exploitative power relations in fieldwork. When I adopt the researcher-as-supplicant role, I expose my reliance on the researched, and emphasize that the knowledge of the researched is greater than that of myself (at least regarding the particular questions being asked). I approach fieldwork as a dialogical process constructed by myself and the researched; the interview becomes an evolving co-authored conversation. The researcher *and* the researched are capable of self-reflexivity and engaging in the self-conscious, critical scrutiny of their multiple subject positions. Like many other feminist geographers, I am influenced by Donna Haraway's (1991) ideas about embodied, partial, situated knowledges. In the context of planning and conducting fieldwork this means that I continually remind myself that they, as the researched, just as I, as the researcher, have partial, embodied, situated knowledges, with the implication being that that we cannot fully know each other.

As I wrote this chapter I mulled over the ideas described in the first two paragraphs of this section in the context of my research about bank managers. Several problems and concerns became clear. First, much of human geography is done *on* the relatively powerless *for* the relatively powerful (Bell, 1978, p. 25). Many of the ethical questions raised by a reflexive examination of power relations in fieldwork really focus on the possibility of exploiting already marginalized people. Of course, as one of those "relatively powerful" people – I am highly educated, white, native English speaking, and have tenure at a PhD granting university – I am also a member of an elite (albeit a less well paid and, perhaps less prestigious elite than that of the managers I study!). Second, the assumption is that the balance of power lies with the researcher, with research strategies often involving efforts to produce polyvocal text, "give voice" or even empower the less powerful. But as Hertz and Imber rhetorically comment "whose purpose does it serve to 'empower' the rich and powerful?" (1995, p. viii). Interviewing elites means dealing with people who, as Linda McDowell found, are "sometimes keen to demonstrate their relative power and knowledge and your relative powerlessness and ignorance" (1998, p. 2137; see also Schoenberger, 1991, 1992; McDowell, 1992c; Herod, 1999). What power do academics hold when interviewing people who are accustomed to having a great deal of control and authority over others? My reliance on the managers was obvious; that the managers' knowledge of women managers in banking is greater than mine was self-evident. And I usually felt I had little control over whether to position myself as a supplicant or not. If anything I was assigned that role by the managers. Indeed, like McDowell (1998, p. 2138), "I did not have to try hard to present myself as an ignoramus" about the world of banking. As with my Columbus project, I was constantly reminded that fieldwork is a discursive process in which the research encounter is structured by the researcher and the researched.

Doing research about elites raises some different sorts of dilemmas, difficulties and concerns than doing research about less powerful, and even vulnerable groups. However, just as with less powerful groups, interviewing elites does involve the purposeful disruption of their lives and can be intrusive. And just as with less powerful people, I am ultimately accountable for my research, for my intrusions into people's lives as well as my representations of the managers in published accounts (Okely, 1992; Stanley and Wise, 1993; Wasserfall, 1993; McDowell, 1998; Parry, 1998). And there still needs to be careful consideration of the consequences of the researcher's interactions with those they research. Subjectivities are contextual, meaning that I (and, of course, the managers) present different personas in accordance with the particular circumstances. So my presence and the managers' response to my presence doubly mediate the interview data I collected from them. Just as I moved between slightly different subject positions with different managers, surely the managers I interviewed similarly choose which version(s) of themselves they wished to present to me (for example, many of the managers referred to the corporate "we").

Moreover, it is entirely possible that they were not entirely truthful with me (why should they be?), that they told me what they thought I wanted to hear, or that they wished to project the right sort of image of themselves as managers and the bank as an exemplary practitioner of employment equity.

Beyond the Interview?

I always tape record interviews and as a general rule, I try to transcribe interviews as soon as possible afterwards. I try to get a transcript to the managers as quickly as I can (although I am frequently less timely that I would like to be). There are several reasons why I think sending a transcript is a good idea. First, I want to be courteous and I especially want to maintain the lines of communication (having worked so hard to open them). Second, (some) things move much faster in the corporate world than in the academic world. So I hope a fast turnaround on the transcriptions will help counterbalance their palpable disbelief at the length of time it would take for anything based on their interview to actually be published in an academic journal ("It'll be ancient history by then," one manager remarked). Third, the managers can provide clarification and feedback. I do not particularly intend this practice to be an attempt to make my published text polyvocal as such. For instance, I have no intention of substituting the grammatically "corrected" text that one manager returned to me. (I assume that seeing his words as he spoke them written down on a page clashed with the polished image he wished to project.)

I am privileged enough that I am able to pay a professional transcriber to transcribe my interview tapes. Some researchers think that the researcher can only fully immerse themselves in "the data" if they transcribe it themselves (see for example Psathas, 1995). In my case, I chose to use a professional transcriber (Mary) for the sake of speed and because, quite frankly, I really dislike transcribing (I spent many, many hours transcribing very long interviews for my PhD research and I fear I may have been put off for life). However, once transcribed I spend a great deal of time revising the transcription while listening to the audio tape. I do this for three reasons. First, to determine what Blake Poland describes as "the trust-worthiness of the transcription" (2001). I listen carefully to the tapes and correct what I perceive to be Mary's misinterpretations of what was said and fill in the blanks representing utterances unintelligible to her. In some cases, the sound quality of my tapes was poor, but also because I was there I am able to reconstruct the conversation. Mary, of course, could not do this. In addition, Mary is obviously not a banker (not that I am, but I have come to understand some of their jargon) and is American, whereas my respondents are Canadian, so there were certain idioms and terms that she did not understand, and had misinterpreted.

The second reason relates to representations of the data on the tapes. I want the transcriptions to represent more than merely the utterances of the respondents, I want to capture a sense of the way the words were spoken. I add notations about (some of) the interactions between the researched and me (and in those cases where two people where interviewed, the interaction between them). I want to add symbols for things like laughter, coughs, intonation, pauses (long or short) and interruptions. (I use some of George Psathas's (1995) scheme for conversation analysis.)

Finally, I consider the process of transcription itself as analysis since it involves the interpretation of the original discussion, including decisions about what to include and what to leave out, as well as where to begin and end sentences. (Customs of language and speaking, for example run-on sentences, do not always translate easily into a written representation of the conversation.) There is not a one-to-one correspondence between the tape recording and the transcription. In other words, the transcript is one representation of the interview. In fact Mary's description of her job: "transcribing is about placing a context on your best guess at what someone is saying, as much as it is about getting the exact words" is also a good description of my job of analyzing and interpreting the interview. I listened carefully to the tapes, engaging in what George Psathas and Timothy Anderson (1990) refer to as "methodical listenings" which are an important part of the analysis because "the status of the transcript remains that of 'merely' being a representation of the actual interaction – i.e. it is not the interaction and it is not the 'data'" (1990, p. 77).

Only occasionally have my experiences in the field matched up with what I anticipated, hoped and planned for in my office. The field always surprises me and it is always changing. As a researcher I find myself constantly maneuvering around unexpected circumstances. This seems to be especially the case when the researched are elites. "Respondents" have not responded to my letters, have not returned telephone calls, lost my surveys, cancelled appointments, cut interviews short, lost interest in my research, withdrawn from the project and so on – in short they have displayed all the usual failings of humans! I have become a "shameless eclectic" and adopted a methodologically opportunistic approach to fieldwork because for me the practice of fieldwork seems to inevitably become the art of the possible and making the best of what you get, rather than the ideal hoped for when planning the research. Also a flexible, opportunistic approach makes me more open to any challenges to my theoretical position that fieldwork almost inevitably raises. For example, I often fine-tune my methodology and even my research question in the field. So almost out of necessity, then,

I have employed mixed methods, triangulating different data sets, and practiced the imperfect art of shameless opportunism in the field.

RESEARCH TIP

Focus Group Research

Preparing for the focus group

- Do a practice focus group.
- Prepare materials, e.g. flip charts, pen, recording equipment.
- Set up and test recording equipment prior to focus group while considering group size, acoustics of room, lighting, background.
- Think about venue ambiance.
- Be generous in your allotment of time.
- Schedule in break times.
- Enable an inviting and supportive environment by providing, for example, refreshments, childcare, and transportation subsidies.
- Think of ways to hear everyone speak, e.g. talking sticks, time limits, speaker's lists.

At the focus group

- As researcher and as facilitator, introduce yourself and explain your research.
- Explain rules of engagement and disengagement.
- Facilitate introductions of participants.
- Acknowledge contributions of participants.
- Follow up and express appreciation for participation.
- Ensure a successful focus group through careful preparation and planning.
- As a researcher, clarify your goals and delineate the expectations of focus group participants.
- As facilitator, explicitly acknowledge tension among participants and work toward a resolution.

13

Studying Immigrants in Focus Groups

Geraldine Pratt

When I collaborated on a research project with the Philippine Women Center in 1995, we chose focus groups as our methodology for studying domestic workers' experiences. Our strategy was to break participating domestic workers into three groups of four to six women each. Well into the process, on the second full day of focus group discussions, I joined a group of four women with whom I had not yet spent time alone, apart from Center organizers. At one point, a woman turned to me and asked, "How about you, Gerry, do you have a nanny?" I told her no, and she noted "That's a personal question." Another offered: "Do you want a nanny? Take me as your nanny." I answered that my child went to daycare. With a joking, but pointed appraisal, a third woman judged: "She will be one of the good employers." The first studied me and said: "You can see in her facial expression. My employer is good but a little bit Tupperware [plastic]." Much was accomplished in this short exchange: the research gaze was turned on me as a potential employer, the focus of questioning was temporarily altered, and a momentary crisis in the possibility of our relationship was voiced and resolved. Declaring me as non-Tupperware signified to me a willingness to continue as research collaborators.

This exchange offers an entry into a discussion of the attractions of focus group methodology for feminist researchers. One attraction is precisely that the relations between the researcher and researched are more open and ambiguous than in one-on-one interviews or survey research. The recent popularity of focus groups arises, in part, out of concerns about hierarchical power relations between researcher and researched, and the limitations of predetermined, closed-ended survey and interview techniques. In the safety of their group, domestic workers could investigate my personal politics and, as a "non-directive" methodology, the focus group afforded considerable freedom to direct the discussion to topics that were important to them and to interpret the researchers' topics in their own terms.

The attractions of focus groups go beyond this, and I elaborate them in the first part of this chapter through detailed examples, mostly drawn from a research project on immigrant settlement in Surrey, an outer suburb of Vancouver, Canada. In this project, separate focus groups were held with service providers, recent Indo-Canadian immigrants, a multi-cultural group of recent women immigrants to Surrey, and second-generation Indo-Canadians in their twenties. The process was also carried out by collaborating researchers (Gillian Creese, Isabel Dyck, Dan Hiebert, Tom Hutton, David Ley, and Arlene McLaren) in four other areas in Vancouver. I thank Dan Hiebert for participating in the service providers focus group in Surrey, and Margaret Walton Roberts and Wendy Mendes Crabb for assistance in setting up and conducting the focus groups in Surrey. (For a discussion of some of the substantive themes that emerged from the focus groups, see Hiebert, Creese, Dyck, Ley, McLaren and Pratt, 1998.)

This chapter is not a manual for doing focus groups. Several such manuals already exist, as for example, Krueger (1994), Morgan (1997, 1998) and Barbour and Kitzinger (1999). Rather, this paper takes up Wilkinson's concern (1999) that feminists have yet to fully realize the potentials of focus group methodology; my hope is that a close reading of several focus group transcripts will encourage feminist researchers to engage these potentials and to think more creatively about the particularity of the information produced through focus groups. Developing a specific analysis of the potentials of focus groups also involves considering their limitations; in the second half of the paper I turn to some of the challenges that I have encountered running focus groups among recent immigrants to Vancouver.

Focus Groups as Process

Focus groups potentially offer a safe space – literally safety in numbers – in which to discuss issues and experiences, and one in which the authority of the researcher can be challenged and negotiated. They also assume and produce a less individualistic mode of knowledge production. Focus group methodology is premised on the notion that we develop knowledge in context and in relation to others. One of the claims that is made of focus groups is that they provide an opportunity to observe directly the process of meaning generation: "how opinions are formed, expressed and (sometimes) modified within the context of discussion and debate with others" (Wilkinson, 1999, p. 67). In interviews, individuals tell us how they would behave or have behaved in certain circumstances; the promise of focus groups is that they provide a setting in which we observe how people behave and make sense of their world in relation to others. Those who recognize this distinction between interviews and focus groups have tended to be disappointed by existing analyses of focus group transcripts, arguing that the processual aspect of focus groups is typically ignored: researchers pay little attention to group interaction, tend to abstract attitudes from discussions, and focus on the content of particular individuals' statements rather than the process of meaning creation (Myers and Macnaghten, 1999; Wilkinson, 1999).

I want to present two segments of transcripts from focus groups in Surrey, with an eye to what emerges through the interaction between focus group participants. The first involves a brief extract from an extended exchange between two service providers who had not met previously, a Chinese–Canadian man, and an Indo-Canadian woman. The sociality of the process is evident, and as the two individuals find common ground through an analysis of their organizations, they develop a critique of Vancouver-based service organizations. Eventually (although not included here) this extends to a critical analysis of the way that the federal government regulates non-profit service organizations.

Dan: So, does an immigrant arriving in Surrey have approximately the same scope for assistance as an immigrant arriving in East Vancouver?

Male Service Provider (MSP): Vancouver has more services for refugees.

Female Service Provider (FSP): I would say that Vancouver societies, if I am not wrong, they are more mainstream, for example, [names agency]. So, as compared to [the agencies of MSP and FSP], we are more community based. Do you find the difference?

MSP: Well, [the Vancouver agency] is close to us. They are similar to [his agency]. . . . Possibly they are both big enough that people find it a little difficult to get to where they want, where they need. It's a matter of them learning about the system.

FSP: I would say that the community organizations over here [in Surrey] . . . and I have seen yours and mine . . . I would say they are more flexible because most of the staff members have gone through the same experience. If somebody's late: five minutes, ten minutes, we say that the person has the right.

MSP: We are more in touch with the community.

FSP: Community. Yes. We are more in touch. We identify ourselves more with the community, and community needs. I am talking on behalf of my society. Just a small example. We have Job Seekers Club programs. Three-week programs. But given their needs, we allow our clients to come another six months, providing extra services until they get a job. That I find a bit different from [the above-mentioned Vancouver organization]. Once the program is over, they stop. But over here, looking into the community needs, we are going deep down into the community. We know their needs more. So then we are more flexible. . . .

MSP: I think I tend to agree. We . . . tend to go the extra mile.

FSP: Extra mile. We go the extra mile.

At first, MSP aligns his agency with the Vancouver one but, as the conversation develops, he refines his position, agreeing that the two agencies in Surrey are more community based, flexible and accommodating. They come to this common position through a process of repeating the same words (community, extra mile) and affirming each other's position.

The next transcript segment does not work in the same way but the formation of opinion through the medium of the focus group is just as evident. The transcript comes from a focus group of recent Indo-Canadian immigrants, one that was in fact organized with the help of FSP and took place at her service organization. The exchange involves two Indo-Canadian men, one (M1) is older and of higher status. I first interpreted this heated exchange as an instance of the censoring that can go on at the public occasion of a focus group. I understood M2 to be censoring the issue of racism. I now think that the exchange is more complex, and that it has little to do with diverting the conversation from the issue of racism in Canada. The disagreement reflects and allows M2 to articulate his criticism of multiculturalism as a mechanism for the classification and segregation of individualism by ethnicity. I quote the transcript at length; this is necessary to convey the movement within the discussion.

M1:   And secondly, you feel more comfortable in a particular community you belong to. For example, if you stay in an area where people from your community are not there, you feel like an intruder. Some of those people are not polite sometimes. Sometimes I am going to Vancouver downtown. I have heard people shouting at me: "You Turban! Go back to India!"

FSP: No! Have you?

M1:   Many times.

Gerry: In Downtown Vancouver?

M1:   Yeah.

M2:   I would like to oppose that.

M1:   I am . . .

M2:   My experience is that . . .

M1:   Wait a minute. I have my experience. I am telling you my experience.

M2:   Yeah, that's okay. No, I just told . . .

M1:   Depends on. . . . No, not many. You don't have, er, you don't encounter certain people.

FSP: It's part of Vancouver.

M1:   Actually, it depends on the time, place for example.

M3:   Sometimes it happens.

M1:   Yeah. [*Others agree.*] It depends on the time. Firstly, in the evening time, when all those drunk people are there.

F1:   Teenagers or something.

M1:   At the same time, certain other people stopped by, and said "Not worth it. He's drunk." I have seen that consolation also. This is, I am telling you my experience. You don't have to agree with it or not.

M2:   No, I don't say that.

M1:   I have not told that all people are like [this]. Certain people at certain times do behave like I am an intruder.

M2:   No. Like I would like to only say my. . . . I am not saying that you don't experience that. But my feeling is that these things may happen. But there it is. I can tell you, you go to Delhi and you are from Punjab . . .

M1:   I also say that I am telling you an experience. Please let me go.

M2:   Okay.

M1:   This experience is not a generalization. This is, at certain times, certain places, for example, on Granville Street in the evenings. I did go. Some people did shout. I am referring to [the fact that] if you are in a homogenous atmosphere, you are not felt like an intruder. [*Tells example from Punjab.*] It's not against the culture here. People are happy. In downtown Granville, when somebody shouted at me, other people stopped. He said "Don't bother. He's foolish." So I am telling the other side also at the same time.

M3:   Yes, segregation is there.

FSP: Segregation is there.

M1:   For example . . . I can give one more. Let's say immigrants come in bundles. More people in one particular area. Immigrants have come. The other people, let's say, I'll call, I don't normally call them whites . . . the, or the . . .

M2:   Euro-Canadians, Euro-Canadians.

M1:   Euroasian people. They move out because the Indians have come here. So they don't feel like, not a homogenous group. They moved out. That's why most of the immigrants are here in Surrey. When an influx of immigrants came to this side, the white people moved out from the area.

Gerry: In Surrey?

M1: In Surrey, yes. People are moving. In an area where more Indians come, the whites move out because they don't feel like staying in that area.

M3: Maybe feeling segregated. [*General laughter.*]

M1: So this study's aim is to? Integration of . . . So you will be sending proposals about segregation? [*Short discussion in which M1 remarks: "Good thing this. Very good topic. In need of this." And then M2 draws us back to earlier discussion.*]

M2: No, I am . . . this . . . I am only asking you a question. I don't understand that. I mean everywhere you see people treat ethnic groups . . . group wise. Why don't they make it like respect the other individual. Forget what his. . . . Suppose I respect you. I don't bother what your other personal life is. Here I feel its more like a group-wide treatment. Like here immigrants. That's also the group. . . . I mean, why is it group wise? I mean I am asking you that. Why . . . I mean in schools and anywhere. It's mostly like they treat as groups. Group as in, "Okay, tolerate this group. Tolerate that group." I mean, why not respect the individual? . . .

F1: In schools?

M2: I am not able to make myself clear.

Gerry: I am just wondering . . . is it something about the way that multiculturalism . . .

M2: Ah, yeah, multiculturalism. I mean multicultural. What is it? Multiculturalism is: you have made the groups one culture. . . . What my feeling is, it becomes . . . then it is a group. So it gives in my mind: "Look at that group" That group, right? Multiculturalism. Okay. Indo-Canadian. One group. Rather than one man. That's it. I don't know to which one you [Pratt] belong to. You are Canadian. That's it. [*Ends.*]

FSP: I think for policy issues, they have to make some categorizations.

M2: It's like in India. Also they have made categorizations. Like we have some people in older times. Untouchable people and all those things. When you start treating them as untouchables, you bring them together. You made a group. Then other groups oppose that group, rather than individuals. Then the group becomes the target.

M1: It's that particular groups have got different needs. That's why they have to grade it.

FSP: Resources are different, you know.

M2: That's okay. But when you say tolerate, the word tolerate. It's like, "Okay, I tolerate you. Stand there." [*General laughter.*]

M1:   Anyway, that's a different story.

M2:   Respecting is something else.

Gerry: Yes.

M1:   Anything else you like to know?

Gerry: I think that's it. Is there anything else you would like to know? [*Silence.*] It's hot too. I can feel it. [*Laughter.*]

M2:   Maybe more than that? . . . I can only say it was a good discussion. Maybe we can thank you for allowing us to blow out some of the . . .

FSP: Our frustrations, our anger, our tensions. [*General laughter.*]

M2:   I like that, see.

FSP: Now that we have given all these frustrations to you, now you look after them! [*Laughter.*]

M2:   It's your responsibility now.

Gerry: Well, thank you very much!

This is a complex conversational event, which begins with M1 explaining his discomfort when he moves out of the ethnic enclave in Surrey. M2's resistance to this perspective causes M1 to qualify his statements about exposure to racist comments in downtown Vancouver, while M1's refusal to let M2 speak forces M2 to expand his analysis beyond an argument against spatial isolation to a fuller criticism of classification by ethnicity. Drawing parallels to the caste system in India, M2 comes full circle and cautions about the way in which these classifications can lead to the victimization of groups of people, a victimization that M1 seeks to avoid by staying within these social and spatial boundaries. M1 also draws on and finds support in FSP's bureaucratic understanding of the necessity of ethnic classification for efficient administration of resources. The focus group provided a space to argue, and through this argument we see complex and important ideas being deployed, for example, the bureaucratic discourse being aligned with a defensive approach to the ethnic enclave. But it is also unclear whether M1's persistent unwillingness to let M2 voice his reticence about generalizing racist experiences in downtown Vancouver causes M2 to express more strongly individualistic feelings than he might otherwise; it is M2, after all, who produces the category of Euro-Canadian when M1 searches for a category other than "white." The arguments of individuals, then, are situated and have to be assessed within the context of others' views.

Wilkinson (1999) interprets this contextual aspect of focus groups as congenial to a feminist perspective; it also moves us toward an understanding developed by Visweswaran (1994) through her historical ethnography of women's involvement in India's independence movement. Through a series of triangulated conversations, Visweswaran discovers that she has been deceived by a number of key informants. In trying to understand the lies and betrayals that led to these discoveries, she moves her attention from a focus on the truth or falsity of informants' statements to the issue of why she is being told particular statements in particular circumstances.

This suggests a fuller contextual engagement than I have thus far suggested, and a peculiarly geographical one. It indicates the need to be attentive to the sites in which focus groups (and interviews) are held, and the value of multi-site analyses to assess the contextuality of discourse. This goes beyond concerns about site selection expressed in how-to-do focus groups texts (that one find a comfortable, accessible, private and neutral space); I am urging geographers to more fully explore the geography of discourse, the ways that focus group conversations change depending on where they occur. In a study of seven-to-eleven year-old children, for example, Green and Hart (1999) find that the rules of discussion vary depending on whether the focus group is held in a school or play group. The researcher's role was more ambiguous and more open to negotiation in the play group; within the school the researcher was firmly situated as teacher. The play group conversations were less orderly and rule-bound, and more extreme stories of risk-taking emerged. This is not to say that the focus groups conducted in play groups were more accurate or more truthful (indeed, there may have been more bravado displayed there), but that they differed depending on the formality of the institutional context. The geography of story-telling, and of opinion formation and expression within focus groups deserves far more attention.

Focus groups are processual in the sense that they offer a glimpse into social interactions; there is a further sense in which focus groups may be useful to a feminist commitment to the process of social transformation: conversations in groups can serve the purpose of consciousness-raising and devising plans for action. I think that Baker and Hinton (1999) are right in cautioning that focus groups do not inevitably lead to participatory action research and that change comes through additional practical acts. Nevertheless, when individuals are brought together in a focus group they can discover shared experiences. This offers support and can lead to an analysis of structures that condition these common experiences, provide opportunities to share information and strategies, and devise new ones. In the focus group with service providers in Surrey, information was shared: for example, the planner in attendance gained knowledge about a multicultural co-ordinating committee of which she was previously unaware. Participants traded knowledge about existing research relevant to immigrant settlement in Surrey and thought collectively about what further research would be of practical use to them. The activist potential of the focus groups with domestic workers at the Philippine Women Center was perhaps clearer. For example, sharing knowledge led one woman to discover that she was not covered by a medical insurance plan, a discovery that precipitated her individual action to remedy this. In telling her story among peers, another domestic worker realized that her employers had paid at least twice her salary for the tasks she does (childcare and housecleaning) when she went on holiday because they had hired two individuals to do them. It was through the telling that this became evident to her. And because these focus groups took place within an activist organization, the transcribed evidence has been used in reports to governments. As a collective activity, the possibility of collective action following from focus group discussions may seem more obvious than in one-on-one interviews.

Challenges

Focus groups offer the possibility of collecting data that shows how ideas are deployed in social interaction in ways that can be put to use for social change; this is not to say that they are without their problems and challenges. Though I have argued that they offer the potential for less hierarchical relationships between researcher and researched, other hierarchies are ever present. These must be carefully negotiated within the focus group, and assessed when interpreting focus group evidence. If we return to the transcript from the focus group with recent Indo-Canadian immigrants, it is clear that M1 is exerting the privileges of age and status. He repeatedly claims his right to tell of his experience, remarks upon the utility of the study, tells M2 when he is off-topic ("Anyway, that's a different story"), and declares when the focus group is over ("Anything else you like to know"?). Returning to the focus group transcript, it is fascinating to see how quickly he establishes his dominance. He is the first to speak. He quickly establishes that what distinguishes Indian from Canadian society is respect for elders. When FSP invites a young non-English speaking woman to participate, M1 establishes that he, rather than FSP, will translate. The following transpires:

M1:   I will translate.

F2:   [*Speaks quietly in Punjabi.*]

M1:   She says she has been here for four years, and she came on married basis.

F2:   [*Speaks quietly in Punjabi.*]

M1: She first came to Kelowna, where her husband was. And later they came to search for jobs here. In search of jobs . . . We can ask her questions. [*M1 asks her a question in Punjabi.*]

F2: [*Speaks quietly in Punjabi.*]

M1: They didn't have any problems regarding housing. They found a house.

Gerry: How? Through friends or relatives? [*M1 and FSP ask her questions in Punjabi.*]

F2: [*Speaks quietly in Punjabi.*]

M1: From, err, through their relatives. They got information and they got contacts. And, err . . . [*M1 asks question, FSP clarifies and F2 provides answer that elicits laughter.*]

M1: Let's elaborate a bit. She, um, gradually, she says she doesn't have any good remarks. [*General laughter.*] But we like to elaborate. [*M1 and FSP ask her questions in Punjabi.*]

F2: [*Elaborates in Punjabi, speaking a little louder.*]

M1: She says she has got frustrated because she couldn't attend school. Because it was too costly. And, um [*asks question in Punjabi*].

F2: No.

M1: She wanted to further her studies. She could not pursue those.

F2: [*Speaks in Punjabi.*]

M1: Yeah, she had a young kid. Nobody was there to look after that kid. So she has to babysit for him. That's why she couldn't go for her studies. [*Asks question in Punjabi.*]

F2: [*Answers in Punjabi.*]

M1: She was in last year of graduation. So pretty well educated. She came from India. So she got married. She got involved with family life. So what else? [*FSP asks question in Punjabi to which F2 answers, laughing.*]

M1: She says when she was in India, she had heard about Canada's garden side, or better side.

Gerry: What about Canada?

M1: Good side of Canada.

FSP: Very bright picture.

M1: Good stories about Canada. The way of life, mobility, no scarcity . . . In India, for example, there . . . the light service goes out, failure of light system. [*General laughter.*] Here, electricity doesn't fail. India, there's no day when it just . . . [*General agreement from the group.*] So these were good things. So she was very pleased to come.

FSP: But over here . . .

M1: Other difficulties sprout here. And she is not happy. What else questions?

Gerry: Because of the education, and help with caring for her child?

M1: Yeah. She has said that due to child care, she had to baby-sit, she didn't have time and err, financial . . . [*Asks F2 a question in Punjabi.*]

M1: She has no financial problems. [*Further discussion in Punjabi.*] Okay, I have a comment for you. Here, my daughter when she came here, she was having Masters degree in Science in India. She was good, brilliant. She came here, married here. She had child here. Then they decided, she and her husband, since you can study, go in for medicine. She applied for medicine. She took the entrance exam. She was top of the list. And after four and a half years she became a full-fledged doctor. So that way, she has now two kids by the time she is a doctor. She has two kids. That's why actually we came in the picture. We had to look after those kids. [*Laughter.*] You can still [*translates for F2 in Punjabi*] study here up to any age. It is no . . .

FSP: Every family has different circumstances.

M1: Yeah. But no financially, she doesn't have problems. So when you are to continue, you can continue. [*General laughter.*] So any other question you would like to ask her?

Gerry: What kind of child care would help now? [*M1 asks question and F2 answers in Punjabi.*]

M1: You don't have father-in-law? Mother-in-law?

F2: [*Answers in Punjabi.*]

M1: She doesn't have any arrangement with any family to look after those kids, so [*asks question in Punjabi*]. Suitable . . .

FSP: But, you know, it is not easy to pay. I think $400 a month for day care. Every family can't afford this much.

M1: So, since you are focusing on integration of immigrants, I think this point can forcibly be put in – that if child care system is adapted to situation like her, then they can pursue their aims and studies . . . childcare arrangements suitable to them. The childcare system are already prevalent here but will not be suitable to her or me, let's say, because of high cost, secondly, their method of handling.

Gerry: Could you talk about that, the method of handling?

M2: [*To M1*] You are putting on your own answers.

M1: No, I'll tell them. I've got to . . .

M2: No. You suggested them, rather than her. The way you want the system changed. I mean, what is the problem handling, what is the problem handling?

M1:   Firstly, the children.

FSP: If you don't mind ... [*Discussion between FSP, M1 and M2 in Punjabi.*]

M2:   [*Pointing at Pratt*] She is the boss!

M1:   You are the boss.

Eventually we ascertained that finances were not the issue, but that F2 had been upset by [an] Indo-Canadian woman in Kelowna who babysat her own child along with F2's child; she judged that this babysitter treated the children unevenly. One pressing issue here is the difficulty of working in translation, to which I return below, but another is the tendency for young women to be spoken for. "Listen" to what happened when another woman, who could speak English, is introduced.

M1:   [*FSP looks to F3*] She came five years ago.

FSP: She is taking ESL classes. She should try with it.

M2:   Yes, she should try.

F1:   Yes, she should try. It's informal, yeah?

F3:   [*With fluency.*] Okay, I came here five years ago. August 1992. I married in August 1992. And I came here. My husband live here and he sponsored me. I got baby in 1993. I got educated, no I . . .

FSP: No. You got less education. Because she has tenth grade.

F3:   I studied in Afghanistan. I came from Afghanistan. And I study here to tenth grade. And after that, like my parents are strict . . . and don't want me to go out alone like that. So I didn't go anywhere in India. Like study more. I came here and I stay at home. I don't have a babysitter. I didn't go back to school for two years. Two and half years. And I came here to [the agency] and I did volunteer work for a couple of hours.

Gerry: Why did you start coming here?

F3:   Because I came here for ESL classes. My husband is working part-time, like night shift. So he take care of my baby. And then I came for ESL classes. And then she told me to . . .

FSP: Apply for a daycare bursary.

F3:   I applied there. And I got the subsidy and now he is going to a babysitter and I'm . . .

This is the last time we heard from F3 although three other people (FSP, M1 and M3) describe her great success in some detail. (M1 describes her progress as a miracle.)

Status hierarchies do not disappear in focus groups and Michell (1999, p. 45) cautions "against a headlong rush into adopting focus groups in an unreflective way if this means further disenfranchising those at the bottom of the social hierarchy." Certainly, as Michell notes, sensitive group composition (in above case, a women-only group – although note FSP's role in speaking for F3) can help. But for some people – particularly stigmatized ones – being brought together as a group can be further stigmatizing. Michell writes from the perspective of a study of eleven-to-thirteen year-old boys and girls. For the lower status girls in particular, there was a marked difference between what they said in focus groups (basically very little) and personal interviews. It was only in the latter that stories of extremely stressful circumstances, particularly at home, came out. She asks that we consider which voices might be silenced in a focus group setting.

Some circumstances may dictate the privacy of an interview. Michell argues that the distinction between the privacy of the interview and the public nature of the focus group is particularly acute when the participants of the focus group are drawn from a relatively closed social world, as was the case in her study conducted within schools. Any statements made within such groups surely circulate beyond them. In these cases, focus groups have the potential of being exploitative because participants are persuaded by the artificiality of the context to reveal intimate details in front of peers with whom they will interact long after the research is over and the researcher is gone (Green and Hart, 1999).

But all focus groups are public performances and, even among strangers, it is unclear where the line between private and public should be drawn. In the focus group with recent Indo-Canadian immigrants, I began to feel that questioning F2 about her childcare was intrusive, especially given the confusion about her financial circumstances and the invidious comparison to M1's own daughter. But if some participants seem vulnerable to over-exposure, the other part of the public nature of focus group performances is that disclosure is selective. From the segments of transcripts quoted, one gets a sense of M1's representation of his daughter. But his glowing presentation of his daughter was typical of every parent in the room: all of their children were the brightest, the most brilliant in their class. Certainly, one of the parents did not reveal that part of their child's pleasure in coming to Canada is that they can live their homosexuality more comfortably, an insight communicated to us by their "child" in a subsequent interview. At the focus group we heard only about their satisfaction with the educational system and success at university. The academic success is beyond dispute, but my point is that disclosure at the focus group was selective. We are now well acquainted with the idea that all social identity is performative (Butler, 1990) but it may be that it is more self-consciously, more cautiously performed on the public occasion of a focus group, especially when participants are acquaintances within a circumscribed community.

A final challenge is also clearly evident in the quoted transcript from the Indo-Canadian immigrant focus group: that is the difficulty of working in translation. It is clear from M2's protests that the translation provided by M1 was less than faithful. M1 also gently subverts F2's concerns about childcare, trying to comfort her with the possibility of returning to her education when her children are older. M1 is no villain here; he simply instantiates a tendency. The same tendency was apparent in a multicultural women-only focus group in Surrey; this case involves a woman translating for four Spanish-speaking women. When a woman from Mexico expressed concern that her son was learning Punjabi because they live in a predominantly Indo-Canadian neighbourhood, the translator – who was not a recent immigrant – responded in the following way: "I am telling her, 'Don't see it as a liability. See it as an advantage. The kid is going to have another language.'" The Mexican woman responded and the translator noted: "Plus she thinks it is difficult for her child to learn two languages. And when they are learning two, I don't think children learn very quickly. Mine speak three." This intermingling of translator's and participants' experiences in ways that undermine those of the participants no doubt reflects the inexperience of the translators involved, as may a second concern: the bland nature of the translated responses. Reflecting on their experiences working with bilingual moderators, Baker and Hinton (1999, p. 105) note the disjointed nature of discussions that are interrupted by the need for an interpreter to translate, and describe the outcome as "matter-of-fact and prosaic responses" in which "emotions and feelings are edited out." Certainly, my experience of working with the Philippine Women Center has been my most successful experience of working in translation. This was because the bilingual moderators shared a culture and had an established rapport with the focus group participants. In this project, we defined our objectives together in English, but the focus groups were held in Tagalog (or a mixture of Tagalog and English for the ones in which I participated). The tapes were transcribed in Tagalog and then translated to English. Acknowledging the problems that attend any translation, this method kept the richness of emotion and detail (more or less) intact.

Recounting Potentials

I hope that I have conveyed some of the potentials of focus groups. I have used focus groups at different moments in the research process. For the Surrey study, they were used in the first stage of research, to get a sense of what research would be useful to service providers and community groups, and to identify key themes and concerns among recent migrants to Vancouver, themes that we pursued in more depth through longitudinal interviews with a small number of households (some of which were recruited from focus groups), as well as a broadly based community survey. In the domestic worker study, the focus groups were the main source of data and I have collaborated with the Philippine Women Center on a number of papers and reports (Pratt, in collaboration with the Philippine Women Center, 1998, 1999; Pratt 1999).

I hope as well to generate some enthusiasm for the *unique* potential of focus groups. They offer something other than a way of collecting interview material efficiently (six to ten interviews at once!). They offer a vantage point from which to observe social interactions. By bringing people together to share experiences, the researcher creates a small group that may generate ideas and plans for action. And although I have not discussed in-depth the "how-to-do" aspect of focus groups, part of appreciating the particularity of focus groups is recognizing that it is a methodology, and not just a loosely organized conversational event. Careful thought must be given to sample selection (its appropriateness determined in relation to the research question), and the moderator must be trained to probe effectively and to lay down the rules that create a permissive environment for fair and open discussion. The focus group is a single event and not equivalent to interviews with the same number of individuals. This means that a single focus group is insufficient; at least three or four should be done to evaluate the consistency of the emerging themes. Krueger (1994) suggests the principle of "theoretical saturation" (Glaser and Strauss, 1967) to determine sample size. That is, you conduct focus groups until the themes stabilize and no new significant information is obtained. Focus groups are public performances, and although interviews are no less performative, they are certainly more private and usually more appropriate for engaging individuals in detailed discussions of their "private" lives. As with any methodology, focus groups are not appropriate for answering all questions; they are extremely useful for sharing experiences and assessing how ideas circulate in a given cultural context.

RESEARCH TIP

Coding

- Maintain a detailed code book.
- Make clear differentiations between small coding units and larger ones.
- Map the data prior to coding, e.g. prepare a set of descriptive statistics for all variables, make notes as to what topics were covered in what part of the interview, point out lacunae in the data.
- Find a method useful for the research project, e.g. colors, letters, numbers, snippets.
- Manage your data by hand or by computer. Software can assist in managing your data once it is coded.
- Persistence and acute observation are the keys to success. For a detailed example of how to approach coding, see Kirby and McKenna (1989).

STUDY MATERIAL FOR DOING FEMINIST RESEARCH

WORDS

Analysis
Autoethnography
Bias
Coding
Eclecticism
Embodied/embodiment
Ethnography
Field work
Focus group
In-depth interview
Multi-method
Open-ended question
Opportunism
Participant observation
Qualitative
Quantitative
Rapport
Research design
Research question
Sampling
Structured interviews
Triangulation

QUESTIONS

Many of the authors note they have difficulty drawing definite conclusions given that their research critiques the certainty of knowledge claims in geography. This resistance is often viewed as a weakness of feminist research. Do you agree with this view? Why or why not?

Some feminist geographers discuss their own complicity in the production of masculinist knowledge in situations such as being a feminist employer, working within the academy, or accepting conditional funding.

How might our research look different if we actually implemented our awareness of our complicity in relations of oppression?

How would you negotiate the following tensions:

Establishing rapport ↔ being self-indulgent
Being strategic ↔ engaging in unethical practice
Doing "relevant" research ↔ pursuing a personal interest
Maintaining confidentiality ↔ creating embodied texts

Explore other tensions between research and activism, theory, and practice.

ENGAGED EXERCISES

Refining research questions through semi-structured interviews

Iteration 1: Set a research question. Find a partner who has similar research interests. Together design two interview guides with between five and seven open-ended questions focusing on the topics of the two research questions. Through this process, each research question is further refined.

Iteration 2: Conduct thirty-minute interviews with one another on audiotape. Be sure that each partner is interviewed about her/his own research topic. Transcribe ten minutes of your own interview. Code it according to examples in a handbook that you find useful. Analyze the data according to particular themes. Themes arising from the data will be useful in refining the research question even further.

Iteration 3: Write up a brief (3–5 pages) interpretation of the data from the interview. Be sure to include a discussion of the research process, the analytical process, and the empirical findings. At the end of this process, the research question should be clarified enough to conduct the research itself. Good luck!

Understanding feminist quantitative research

Read the following article:
Gordon, P., Kumar, A. and Richardson, H. W. 1989: Gender differences in metropolitan travel behaviour. *Regional Studies*, 23, 499–510.

Using a large national sample in the US, this study found that women consistently have shorter work trips than men, regardless of income, occupation, marital and family status, mode of travel, or location and that women undertake more non-work trips than men. The authors, however, concluded that: "we are certain that many previous researchers have jumped too easily from the objective fact of shorter female worktrips to subjective explanations insufficiently supported by evidence but which happened to fit very neatly their feminist preconceptions of sex discrimination with respect to labour markets" (Gordon, Kumar and Richardson, 1989, p. 500).

Critically evaluate the data used, statistical analysis, arguments, interpretation, and presentation of the results in the paper in light of feminist epistemologies and methods. What kind of methodological flaws can you identify? Do you think the authors' conclusion is warranted? Consider the following:

How was women's household responsibility measured?

Who were the individuals studied, and what groupings of these individuals were used in the statistical analysis?

Which geographical area was included by the study? How was location in the regression analysis represented?

SUGGESTED FURTHER READING

Several methods handbooks exist, some of which are more useful to feminists than others. The most wide-ranging reference book on feminist methods is Shulamit Reinharz's (1992) *Feminist Methods in Social Research*. The book covers a wide range of methods that can be effective in feminist research, as for example, surveys, ethnography, cross-cultural research, and content analysis. Although dated, Helen Roberts' (1981) edited collection, *Doing Feminist Research*, is still useful in identifying the dilemmas feminists face in doing research. For a step-by-step account on how to do a feminist research project, Sandra Kirby and Kate McKenna's (1989) *Experience Research Social Change: Methods from the Margins* is the best "how-to" book we've come across. They include a number of techniques that assist at each stage of the research process – from deciding on a topic and research question, to managing and coding qualitative data. Margrit Eichler (1991) provides an overview of ways to excise sexist bias in a range of research methods.

Several topic-specific books are effective in preparing to undertake research. Such topics include oral history (Gluck and Patai, 1991), interviews (Kvale, 1996), community change (Ristock and Pennell, 1997), participatory action research (Maguire, 1987), focus groups (Barbour and Kitzinger, 1999), community activism (Naples, 1998), and auto/biography (Stanley, 1992). Interdisciplinary readers on feminist methodologies and topical handbooks, too, may be useful (Hesse-Biber et al, 1999; Guerrero 1999). Although not necessarily feminist, Celia Gahan and Mike Hannibal (1998) might be useful in using QSR NUD*IST as a qualitative analysis software package.

In addition to these reference books, Sage Publications has a series of methods handbooks that can be helpful in figuring out which methods to choose for which questions, including the *Handbook on Qualitative Research*, edited by Norman K. Denzin and Yvonna S. Lincoln (1994). This enormous handbook is now published in three separate paperback volumes: *The Landscape of Qualitative Research: Theories and Issues* (1998c), *Strategies of Qualitative Inquiry* (1998b), and *Collecting and Interpreting Qualitative Materials* (1998a). When used in tandem with feminist books on methods, methodologies, epistemologies, and research, these resources can be helpful for feminist researchers in geography.

Beyond the feminist use of methods and feminist methodologies, there are issues within the research process that feminists need to consider. For example, Gesa Kirsch (1999) talks about the wide range of ethical quandaries feminists face when undertaking to do research *as* a feminist: subject positionings, interpretations of data, and the politics of publication. Lee Harvey (1990) pulls together a set of research examples and lays out in analytical terms what is critical about each research project. Both Diane Wolf (1996) and Rae Bridgman, Sally Cole and Heather Howard-Bobiwash (1999) edit volumes wherein primarily anthropologists identify difficulties in undertaking feminist research and elaborate the complexities of the politics of feminism within research. Both volumes contain a range of readings that take up the types of issues feminist geographers deal with on an ongoing basis. A special theme issue of *Resources for Feminist Research/ Documentation sur la recherche féministe* (2000, volume 28, nos. 1–2: 9–243) addresses contemporary dilemmas feminists face in a variety of qualitative research, including being a feminist in the academy (Moss and McMahon, 2000), traversing the spaces between fieldwork and the classroom (Johnson, 2000), negotiating institutional parameters within action research (Reid, 2000), and deconstructing the interview (Lyons and Chipperfield, 2000).

Outside this list of references is a plethora of journal articles addressing issues arising out of the research process. A wide range of geography journals publish articles that might be of interest to feminist researchers in geography, such as *Annals of the American Association of Geography, Antipode, Area, Environment and Planning A, Environment and Planning D: Society and Space, Gender, Place and Culture, Geoforum, The Professional Geographer*, and *Transactions of the Institute of British Geographers*, as well as a number of women's studies journals, such as *Atlantis, Feminist Studies, Frontiers, Gender and Society, International Women's Studies Forum*, and *Signs*. Flipping through the pages of recent issues will no doubt open up literature searches about methodological issues in feminist geography.

14

Further Notes on Feminist Research: Embodied Knowledge in Place

Isabel Dyck

The chapters in this book exemplify the notion of feminism as a *process*, as laid out in chapter 1, as a manner of asking questions and a way of looking at the world. They illustrate different concerns for feminist researchers in geography and ways of addressing these as they seek to describe and explain our social and spatial worlds in ways that are inclusive of women positioned variously in relations of power. In showing a variety of ways of putting into practice geographical research that addresses feminist questions and concerns, the authors indicate that how research is done has a specificity related to *where* a study is conceived, designed, and carried out. Space and place matter in conducting studies as well as in constituting and framing the experiences of the women who are the focus of inquiry.

In reflecting on their research experience, the authors articulate the webs of tightly interwoven processes that inform and shape both the constitution of geographical knowledge through feminist research and how that knowledge comes to "count," or hold authority in understanding our social and spatial worlds. The exercise of careful reflexivity, a common theme in the chapters, shows the vagaries of the research process for the researcher – not just in "gathering" data but in establishing research relations, writing, and publishing. The authors' reflections emphasize fluidity and the instability of research relations and boundaries that are hard to predict, but instead unfold during the research process. They give careful attention to moral dilemmas and the positioning of researchers in their research. The discussion over and over again emphasizes the close interconnections between theory, method, and epistemology. Through the discussion, the authors uncover the uncertainties of the knowledge constructed through research, as the local, embodied knowledges of the "field" are translated through interpretive acts into the printed and spoken word that goes on to inform categories and concepts that are given life in other contexts, as for example the policy document or media story. Constructing knowledge from field research is a potentially daunting task and a heavy responsibility when, as several authors suggest, it is from a context of fluidity and uncertainties that we eventually "fix" meaning!

In these concluding notes I pursue some of the issues emerging from such fluidity and uncertainty as they have interested me in my own research. I focus on "context" and power, using the notion of geographic scale to help trace some of the implications of the shifting, continually changing research situations and relations that we encounter as researchers. Pamela Moss emphasizes (chapter 1, this volume) that the practicalities of doing research are embedded in the relationships among contexts, power, and knowledge; in verification the chapters in the book indicate that there is a spatiality to research relations, laden with complex relations of power that are negotiated throughout the research process. Here I aim to draw out some of the issues relating to spatiality and context that I have found useful in thinking through what it means to do feminist geographical research in *specific* places. This means, for me, moving away from prescriptive research "to dos" and thinking about research issues as an embodied researcher whose interpretations tell a particular story of others' lives and embodiment.

Personal and Disciplinary Journeys

It is clear from the chapters of this book that feminist geography has come a long way since the early work of the 1970s. Yet struggles remain, albeit unevenly. Several chapters tell of personal journeys through the challenging terrain of practicing geographical research that is feminist. Some of the challenges in doing research about the ways places and spaces make a difference to how women live relate to responsibilities in how data are used. Who will benefit from our research? How can our analyses go beyond local meaning to make points about structures, discourses, and practices of inequality that are relevant to many women's lives in different parts of the world? Other challenges come from the practical and emotional experiences of doing research in a specific place, whether this is familiar to us as "our" country, "our" city, or carries the unfamiliarity of a perhaps exotic "otherness." As well-rehearsed in feminist methodological literature, much of the challenge of research comes from a feminist sensitivity to issues of power and ethics, not readily resolved (see for example Wolf, 1996). But such challenges are highly contextual; who one is and where one is make a difference to how power plays out and ethics unfold in a particular research project. There are intricate links between power and "context" for they frame what we choose to research, how we approach our research, and what issues emerge as dilemmas in the field.

I have been surprised over the course of several qualitative research projects that new, unanticipated dilemmas and difficulties continue to arise. Experience does come to aid, but still the realities of doing research in the world that we are a part of can throw up previously unencountered issues. I describe a few of these later, but first I want to make the direct relation between such "surprises" and the notion that research is embedded in layers of context, a notion that has been explored particularly in discussion of ethnography (see di Leonardo, 1991, for a useful exposition). Research relationships, the situatedness of the researcher within intellectual currents, and the location of the research and researcher within wider sets of political and economic relations at a particular historical juncture all shape how social and cultural realities are perceived and constructed. While I think many researchers implicitly recognize this layering of context, they less often integrate it into analysis. For example, the fluidity noted in several places throughout the chapters in this book, the authors do not often pursue the notion of layering as an important feature of the meanings made in research about places and people they talk about. Through the notion of scale – moving from state level to where we live and work (our everyday worlds) to the specific setting of research, such as the home of someone we interview or other such "field site" – the importance of that fluidity to how we do research and the dilemmas we may encounter can be grounded in material conditions, specific in time and space. All aspects of research, indeed, are embedded in a layering of contexts. Yet these contexts shift, and in some places and times more rapidly than in others. What research I choose to do, how I do it, and the specific dilemmas I encounter cannot be completely abstracted from the materiality of the "where" I do the research.

As geographers we are in a good position to explore the intertwining of place and subjectivity in processes constitutive of knowledge. Geographers also bring an awareness of the significance of movement to social positionings, identity, and experience – whether of the individual moving in time and over space in a particular place, or of movement of peoples on a world scale as processes of globalization dig deep into local economies and local lives. Massey (1993) captures these connections between the global and the local evocatively in her description of a London neighborhood where people, goods, stores and services reflect larger scale movement of peoples and settlement processes that result in uneven social and cultural transformations. Power is necessarily part of such movement; who moves where and how, what part they play in the global economy as it is experienced as "local," and what access to social, economic and political resources they have are all outcomes circumscribed by the operation of power – again at different scales. Immigration policy, political economy, local labor and housing markets, and everyday social interactions and institutional practices experienced in the course of a day in home, neighborhood and workplace come together to construct identities and experience. Context, power, and fluidity are embodied in those whose lives we study and, just as importantly, in our own.

My experience of movement through disciplines and between geographical locations has formed how I look at the issue of context within research as well as what I view as research dilemmas and issues important to consider in doing feminist geographical research. I have moved from a conservative health science through a politically left social anthropology to feminist geography. In the course of my movement, I shifted from what I first experienced as a "white" industrial town in the UK encapsulating "traditional" norms and ways of living gender, to more diverse cities in the UK, to a divided (predominantly white and First Nations) Canadian prairie city, to a west coast "multicultural" Canadian city that has experienced rapid cultural transformation in only a few decades, and back and forth to that formerly white, industrial town in the UK which now is largely de-industrialized and home to many former non-white colonial subjects of Britain. I have also moved from the "certainty" of a scientific paradigm and the body of medicine to ethnographic and other qualitative approaches with their different standpoints and ways of viewing knowledge. This movement through disciplines, intellectual currents and changing places has emphasized to me the fluidity of places and people, the shifting nature of contexts, the malleability of identities, and the way subjectivities and places are embodied in a fluid way – we are where we are, as well as "who" we are as gendered, classed, "raced," etc. (see Grosz, 1994).

Thinking about the people and places of my personal journey signals to me the importance of geography in understanding the research relations located within a layering of social relations, movement of peoples, fluid subjectivities, and changing place identities. Geography matters at different scales, and one's social positioning and insertion in power relations has a material accompaniment. Geography matters in that it makes a difference to how subjectivities are formed and lived, with regional and national differences coming into play in constituting social, economic, and political life and how people carry out their day-to-day lives. It also matters at the scale of the specific spaces in which people conduct their daily lives; identities are lived through bodily experience in the materiality of those spaces in and across which the body moves over the course of a day. This materiality of the everyday has been an important concern for many feminist researchers. It is from this location that we experience our worlds, our gender, our class, our "race." This is also bodily experience, our body being the medium through which place is lived, with its layering of relations of power.

It is from this point of connection, between body and place, that I now go on to reflect on some of the issues in the practice of doing research opened up in the preceding chapters. Conceptualizing these issues through different scales, helps to reiterate the links among "context," power and fluidity that affect research practice. As the chapters of the book emphasize we as researchers are part of this context. The in-betweenness of the "researcher" and the erasure of the notion of the field, as noted by Nast (1994), is particularly pertinent to thinking through research as an *embodied researcher* interacting with research participants, each with her own history and geography brought to the research. This takes on particular importance when working in the fluidity of many contemporary places which are experiencing rapid social and cultural transformation.

Doing Research "In Place": Context, Bodies and Embodying Knowledge

If we think of context at various scales, we can keep in the forefront of analysis the connection between everyday lives in particular places and institutional practices and relations of power. Funding agencies and their review boards are also important players in doing research, sometimes uncertain as guidelines and priorities change. Favored topic areas may be subject to institutional agendas and political sensitivities, negotiated further by review committees. Feminist methodological frameworks for research, as in critical qualitative research in general, may not readily fit into the micro-accounting and strict time orientations required by some granting agencies. It is perhaps not surprising that a fairly narrow range of methods are used in feminist research in geography and appear in this book. Practical constraints of budget and time, set with institutional expectations of career paths, make it difficult to create a fundable research proposal based on the exploratory nature and community building processes required in developing action research that might be closer to advocated feminist principles of research, where collaboration is key. The politics of power are playing out through the day-to-day practices of governments and institutions, and these inevitably affect at some level how we do our research. These are place-based issues, reflecting national, federal, and university concerns and priorities as the distribution of limited financial resources are organized.

Our relationship to place and how we live this also enters our research in other ways, whether this is research carried out "at home" or "away." I am not pursuing this relationship in terms of the insider/outsider debate that has flourished primarily in anthropology, but wish to look at this in terms of some specific dilemmas that have emerged from my own research with immigrant families in Vancouver in studies conducted with Arlene McLaren, a sociologist. These are in part related to "wider" issues connected with research funding and in part related to our concerns as feminists in doing research in a way we consider ethical as mediated by the conditions of a particular place that has undergone rapid cultural transformation. Canada as a nation has a political commitment to multiculturalism, which aims to afford equal participation in society regardless of cultural identity. As might be anticipated this is not a seamless or unproblematic process, and settlement and integration issues (through the language of social cohesion) merit attention on the nation's agenda, as reflected in a multiply funded project and its four research centers in Canada. Vancouver's experience of Canadian immigration policy change and world population movement is reflected in its rapid demographic change since the mid-1980s that has resulted in a highly visible Asian and, to a lesser extent, South Asian presence in the city and its environs. Neighborhoods and suburbs of the city have changed from their predominantly white complexion over this period, although as would also be anticipated, unevenly. Doing research in one of these suburbs has been a highly "contextualized" experience; funding guidelines include a commitment to community participation and training new researchers (graduate students) and, as researchers, we were continually reminded of our privileged position of whiteness. In studying diversity "at home" the simple binary of "self" and "other" was complicated as we negotiated our identities from a position within a non-homogenous whiteness – with its changing meanings in the everyday living of multiculturalism – with study participants, whose identities within Canada were in flux as they were creating a "new" life in a new "home" country.

Into this complex equation came research assistants, with their own histories and geographies, but with the mandate of conducting in-depth interviews primarily through our – mine and Arlene's – lenses. They included graduate students and non-university based research assistants. All were white; one woman a recent immigrant from Mexico, one a recent graduate, and two current graduate students. All had some, yet widely varying, experience of interviewing. We felt the students and community research assistants were gaining invaluable experience and skills from the project that could be translated into their own graduate student work, or would help the non-student research assistants, one of whom was a recent immigrant, gain employment in a city where immigrant related services and issues are prominent. They were pivotal to the success of the research as intermediaries in the research process, building community links and facilitating recruitment, as well as interviewing. The demands of the research were high, and required research assistants to be considerably flexible in their approach to research as they dealt with complex political and personal dynamics including, for example, sometimes interviewing family members together. While the training element of such work is important, research conducted by people other than the principal research investigators brought ethical issues in using research assistants to the fore. We were concerned with the emotional distress one graduate student research assistant had to address in one interview and the position she was put in as a mother and daughter used her as an intermediary in a dispute between themselves. Another research assistant was upset by the marriage dynamics she witnessed. As social science researchers we rarely have the training to handle complex emotions and relationships and have to think on our feet, as we expected our research assistants to do. We bring our subjectivities to research in different ways, and have responsibilities to research assistants as well as to study participants, yet there is little discussion in the literature about the emotional dimensions and demands of doing research, especially in the context of conducting research through someone else's lenses.

Another issue related to subjectivities in place, is that of the methods used in research. The power *of* ethnographic methods of social anthropology in exploring social and economic life in a holistic way, is twinned with the concerns of power *in* ethnographic approaches leading to questioning of the links between feminist inquiry and ethnography and possible resolutions (Stacey, 1988; Smith, 1990b). In geography, ethnography is not a common approach. More usual is the use of in-depth interviewing which represents a modification of an approach which intends to give space for the previously subjugated (or invisible) knowledge of the "subordinate other." "Studying up" of those in positions of power uses the same approach – although the power relation between researcher and "researched" is then more ambiguous than in interviewing those we consider to be in a similar position to ourselves or those who are in less powerful positions in society. A focus on power differential in the interview, however, needs to be located in those wider layerings of context if the interview is to be understood as more than a simple matter of researcher–"researched" dynamics.

Ann Oakley's (1980) important article on interviewing women opened a floodgate of discussion on power relationships in the interactive interview. Geographers have also added to understanding of the significance of the interview setting, typically the home, or other fieldwork settings to the knowledge we construct (see for example Nagar, 1997; Oberhauser, 1997). Certainly the locales of settings for interviews are ladened with power relationships, as again in the case of my own research these being framed by the differential positionings of immigrant study participants and researchers within the unspoken whiteness of multiculturalism. As places change, so too methods may need to change, including the location in which, for example, the in-depth interview takes place. The "safe" space of the home advocated in much feminist research may not be so safe for the recent immigrant or woman in a vulnerable position. One-on-one interviewing methods, also advocated in feminist research in order that a woman may speak unhindered, may also have to be adapted in the social reality of doing research. In our research experience with newly immigrated women, we have interviewed mothers and daughters together, husband and wife together, sisters together, whole families, and other combinations. Increasingly, some type of focus group or group interview is found to be more amenable to study participants who may be unused to western interviewing, or indeed, tell tales in a different way. Collective story-telling may also be combined with individual interviewing in gaining a more holistic picture of a topic. As contexts shift, intellectually and geographically, a flexible approach to methods is a necessity. And what we consider appropriate feminist methodology may have to be sensitive to the conditions in which we do research – much can be learned from negotiating *how* and *where* stories are told.

The fluidity of the contexts within which we negotiate our research and ourselves precludes a simple approach to accounting for and working with the issues of power that feminists have centered in methodological discussion. Just as we get something "right," another dilemma or puzzle is likely to confront us in the exacting venture of doing ethically sound research that takes our responsibilities as feminists seriously. We may be tempted to blame ourselves for problems in the field, but in doing so miss a chance to bring an analytical lens to such problems which can help us learn more about the links between the practical issues of fieldwork and the context of our research. I have found thinking through the body a useful lens for exploring links between power and knowledge in conducting fieldwork. Feminists drawing on poststructuralist thought have focused on the inscription of normalizing discourses on bodies, such as those of "gender," "class," and "race," set against the "unmarked" heterosexual male, middle-class, and white subject, and how this both mediates and (re)constitutes experiences, identity performance, and spaces through regulatory practices and self-surveillance as the body as "text" is made and "read." In relation to constructing knowledge(s), the body can be viewed primarily in two ways. First is the body of the researcher, as the key research "instrument" or "tool" of qualitative research, and, second, the bodies of research participants – those we study – tell us more than the words of the transcripts of their accounts, and help to integrate the body with other geographical scales. In these ways research is *embodied*. The notion of embodiment emphasizes the material spaces within which and from which women speak or, alternatively, are silenced. It is from embodied positions that threads between situated, local knowledges, and the organizing relations of power can be followed, challenging a macro and micro distinction in analysis. As well, there are various embodiments in the research process to be considered – of the researcher, of those who participate in studies as research subjects, and their respective positions in social relations and discourses through which bodily experience and knowledge are constructed.

Relations between those conducting research and those being "researched" have long been a central focus in feminist methodological debate. We are reminded of the complexity of the power dynamics of such relations throughout feminist work in many disciplines, as well as within geography. We are also reminded that our self-presentation in research is an important dimension of the politics of fieldwork, although we may not always know what may be appropriate to a particular situation beforehand. This "surface embodiment," through "adorning" ourselves in a way we see as consistent with our identity as researcher in the specificity of place and context of a particular study, is one dimension of the body politics of the field, although gendered and "raced" identities are less malleable. I also see the focus on body and its movements in the specificities of space as interesting to think about in understanding how the bodies of both researcher and researched may be used strategically or embody discourses of power in the complex communication processes defined within an interview setting coded as a space for gathering data (the researcher) or providing a personal narrative (the study participant). For example, the resistance of the elderly Punjabi woman who sat cross-armed with eyes shut, declining to answer the questions posed by a research assistant, while her daughter, who encouraged her mother's participation, responds for her. Or the interviewer with her tape recorder, microphone, notes, and arrival by car who brings in the "outside" authority of the university and transforms the private space of the home into an "official" interview site, rearranging the space as outlets for the tape recorder are found or it is set up on a convenient table.

While we may be aware of, and perhaps make a point of recording the bodily comportment and dress of those we interview in our fieldnotes, we can be less certain of how our bodies and their political positioning are "read" by those we interview. In research with immigrant families we felt our positioning as "university researchers" might be read differently. In some instances women clearly had paid careful attention to their appearance and dress for the occasion of the interview which was an opportunity for them to link their concerns to people they perceived may be influential. Did our style emphasize our links with authority-laden institutions – either intimidating or, alternatively, increasing confidence in the importance or legitimacy of our research? But beyond our surface adornment were bodies inscribed with gender and whiteness, and moved and acted in culturally specific ways, symbolizing the different positioning and entitlements in Canadian society that we hold *vis-à-vis* the women we interviewed. In this study we embodied the layered narratives and relations of multiculturalism, a national research initiative on immigration, and the institutional codes and practices of university-based research (Dyck and McLaren, 2001).

Patricia Price (1999) discusses another way of thinking about the body in research. She sees the body as a site of embodied knowledge *and* a scale for the focus of investigation. Interviewing women in Mexico she sees three sites of the "very local" (domestic violence; religious faith; and inner landscapes of hopes, fears and dreams) as entry points to the embodiment of socioeconomic processes and increasing poverty as Mexico has been subject to economic restructuring. Comportment, poor mental health, and other dimensions of corporeality, such as the bruised bodies of domestic violence, *literally* embody the socioeconomic and political relations inscribed upon them. Thinking about the body in such a way in considering interactions between researcher and "researched" can bring an added dimension to our understanding of transcribed text and fieldnotes in which our interpretations are grounded. Inscribed in different ways through discourses and practices of power, we read each others' bodies. The tears, hushed tones, and other signs of distress expressed by some women to us as we talked of their immigration and settlement experience in Canada embodied the struggles, fears, and disappointments as the "promised land" did not always deliver. Work was difficult to find, and women at home were often isolated and lonely. Certainly attention to the body, whether that of the embodied researcher as primary research "agent," or the body of a study participant as a site of knowledge embodying social processes and material practices helps to keep in the forefront of analysis the connections between everyday lives in particular places and institutional practices and relations of power

Moving toward Embodied Research

Feminist scholars have been particularly searching in approaching *how* we know the worlds we live in; identifying and attempting to make apparent subjugated knowledges that have been buried within discursive and material processes and practices that have naturalized women's and men's sexed bodies, gendered activities, use of and movement across space, and organization of power relations that reach every corner of social, economic, and political life. Contemporary feminism's concerns with difference and identity and ongoing discussion of epistemological issues have reiterated the importance of *embodied* knowledge in pursuing a materially grounded feminism. As feminist researchers, we are urged to choose methods that ensure the (bodily) experiences of women are the basis of our analyses, and that the scrutiny we apply to our field methods includes reflection and analysis of the ways *they* are embodied as we translate "data" and create the legitimacy of our interpretations through the social and political processes of academic knowledge construction.

In this volume, the authors provide a number of examples of ways of operationalizing feminist research in geography that implicitly recognizes the centrality of experience through the body as the basis from which to construct knowledge. In addition to being exemplars of "doing gender" *in* the specificities of place, the chapters also show that research is not a matter of taking "abstract" feminist research principles and applying them in any simple way. Ways of doing feminist geographical research have emerged from feminist geographers' experiences in concrete situations that have their own materialities and imbricated pattern of relations of power. However, such local settings and the situated subjectivities of those we study within them need to be placed in processes and discourses beyond the immediate visibility of the interview setting.

A further point arising from these chapters is that the snap-shots of people's lives from which we construct knowledge are located in a narrative flow with which we may have little or no ongoing, direct connection. Our interpretive acts – analysis, writing, and presentation – "fix" such ethnographic moments in the tales that we tell, perhaps with consequences we may not be aware of. While the exercise of reflexivity in our research cannot make all transparent (after Rose, 1997), it is important to continue to make our best efforts to uncover the mechanisms of the truth claims we produce. This is perhaps particularly important when speaking as feminists. Chandra Mohanty (1991) draws our attention to the heterogeneity of feminist discourse and practice, with these grounded in specific cultures, histories, and geographies. White, western feminist influences dominate geographical work at this time. In this volume, the authors, rather than rigidly defining feminist research practice, open up a space for imagining possibilities of doing feminist research in different ways. The examples of research in this volume provide an exciting array of insights into what it is like to do research. Mistakes can, and are, made, lessons learned, and research practices adjusted to engage us in a richer way in our research. As the Feminist Pedagogy Working Group notes (p. 23, this volume), there is no one "good" way to do feminist research – every project must create logical and practical links to approaching a specific topic in a particular context.

References

Abu-Lughod, L. 1988: Fieldwork of a "dutiful" daughter. In S. Altorki and C. F. El-Solh (eds), *Arab Women in the Field*. New York: Syracuse University Press, 139–61.

Acker, J., Barry, K. and Esseveld, J. 1993: Objectivity and truth: problems in doing feminist research. In M. M. Fonow and J. A. Cook (eds), *Beyond Methodology: Feminist Scholarship as Lived Research*. Bloomington: Indiana University Press, 133–53.

Alcoff, L. and Potter, B. 1993: *Feminist Epistemologies*. London and New York: Routledge.

Alton-Lee, A. and Densem, P. 1992: Towards a gender-inclusive school curriculum: changing educational practice. In S. Middleton and A. Jones (eds), *Women and Education in Aotearoa Vol. 2*. Wellington: Bridget Williams Books, 197–220.

Anderson, E. 1995: Feminist epistemology: an interpretation and a defense. *Hypatia*, 10 (3), 50–83.

Anderson, K. and Jack, D. 1991: Learning to listen: interview techniques and analyses. In S. Gluck and D. Patai (eds), *Women's Words: The Feminist Practice of Oral History*, New York: Routledge, 1–26.

Antipode. 1995: Discussion and debate: symposium on feminist participatory research. 27, 71–101.

Atkinson, P. 1991: Supervising the text. *Qualitative Studies in Education*, 4 (2), 161–74.

Babbie, E. R. 1973: *Survey Research Methods*. Belmont, CA: Wadsworth Publishing Company, Inc.

Bailey, C., White, C. and Pain, R. 1999: Evaluating qualitative research: dealing with the tension between "science" and "creativity." *Area*, 31, 169–83.

Baker, R. and Hinton, R. 1999: Do focus groups facilitate meaningful participation in social research? In R. Barbour and J. Kitzinger (eds), *Developing Focus Group Research: Politics, Theory and Practice*, Thousand Oaks, Calif.: Sage, 79–98.

Barbour, R. S. and Kitzinger, J. (eds) 1999: *Developing Focus Group Research: Politics, Theory and Practice*. Thousand Oaks, Calif.: Sage.

Bäschlin, E. and Meier, V. 1995: Feministische Geographie – Spuren einer Bewegung [Feminist Geography – Footprints of a Movement]. *Geographische Rundschau*, 47(4), 248–51.

Behar, R. 1996: *The Vulnerable Observer: Anthropology that Breaks Your Heart.* Boston: Beacon Press.

Bell, D. and Valentine, G. 1995: *Mapping Desire: Geographies of Sexualities.* London: Routledge.

Berg, L. D. 1994: Masculinity, place, and a binary discourse of theory and empirical investigation in the human geography of Aotearoa/New Zealand, *Gender, Place and Culture*, 1, 245–60.

Berg, L. D. 1998: Reading (post)colonial history: masculinity, "race", and rebellious natives in the Waikato, New Zealand, 1863, *Historical Geography*, 26, 101–27.

Berg, L. D. and Kearns, R. 1996: Naming as norming: "race," gender, and the identity politics of naming places in Aotearoa/New Zealand. *Environment and Planning D: Society and Space*, 46, 99–122.

Bernstein, S. 1992: Confessing feminist theory: what's "I" got to do with it? *Hypatia*, 7 (2), 120–47.

Blunt, A. and Rose, G. (eds) 1994: *Writing Women and Space: Colonial and Postcolonial Geographies.* New York and London: The Guilford Press.

Bock, S., Hünlein, U., Klamp, H. and Treske, M. (eds) 1989: *Frauen(t)räume* [*Women's spaces/dreams*], *Urb et regio* Vol. 52. Kassel: Kasseler Schriften zu Geographie und Planung.

Bogdan, R. and Biklen, S. 1992: *Qualitative Research for Education. An Introduction to Theory and Methods.* Boston: Allyn and Bacon.

Bondi, L. 1990a: Feminism, postmodernism and geography: space for women? *Antipode* 22, 156–67.

Bondi, L. 1990b: Progress in geography and gender: Feminism and difference. *Progress in Human Geography*, 14, 438–45.

Bondi, L. 1992: Gender and dichotomy. *Progress in Human Geography*, 16, 98–104.

Bondi, L. 1997: In whose words? On gender identities and writing practices. *Transactions of the Institute of British Geographers*, 22, 245–58.

Bondi, L. 1998: On referees and anonymity: a comment on Richard Symanski and John Pickard's "Rules by which we judge one another." *Progress in Human Geography*, 22, 293–8.

Bördlein, R. 1994: Geographinnen an Hochschulen in der Bundesrepublik Deutschland, Österreich und der Schweiz [Women geographers at universities in Federal Germany, Austria and Switzerland]. Materialien 17. Institut für Kulturgeographie, Stadt-und Regionalforschung. J. W. Goethe-Universität, Frankfurt a. M. Compilation reprinted in *Geographica Helvetica*, 1, 31–40.

Bowles, G. and Klein, R. D. (eds) 1983: *Theories of Women's Studies.* Boston: Routledge and Kegan Paul.

Bridgman, R., Cole, S. and Howard-Bobiwash, H. (eds) 1999: *Feminist Fields: Ethnographic Insights.* Peterborough, Ont.: Broadview Press.

Brod, H. and Kaufman, M. 1994: *Theorizing Masculinities.* Thousand Oaks, Calif.: Sage.

Buff, E. 1978: Die Abwanderung von Frauen aus dem Berggebiet. [Women's emigration out of mountain regions]. Magistra Thesis, University of Zürich, Department of Geography.

Bühler, E., Meyer, H., Reichert, D. and Scheller, A. (eds) 1993: *Ortssuche: Zur Geographie der Geschlechterdifferenz. [Looking for place: To a Geography of Gender Difference]*. Zürich-Dortmund: Schriftenreihe des Vereins Feministische Wissenschaft.

Buschkühl, A., 1984: Die tägliche Mobilität von Frauen: Geschlechtspezifische Determinanten der Verkehrsteilnahme. [Women's Daily Mobility: Gender-Specific Determinants in Traffic Patterns]. Magistra thesis, University of Giessen, Department of Geography.

Butler, J. 1990: *Gender Trouble: Feminism and the Subversion of Identity*. London and New York: Routledge.

Buttimer, A. 1990: Geography, humanism, and global concern. *Annals, Association of American Geographers*, 80, 1–33.

Buttimer, A. 1993: *Geography and the Human Spirit*. Baltimore: Johns Hopkins University Press.

Canadian Geographer. 1993: Feminism as method, 37, 48–61.

Cassell, J. 1988: The relationship of observer to observed when studying up. In R. G. Burgess (ed.), *Studies in Qualitative Methodology, Vol. 1*. Greenwich, Conn.: JAI Press, 89–108.

Chouinard, V. 1997: Structure and agency: contested concepts in human geography. *Canadian Geographer*, 41, 363–77.

Chouinard, V. and Grant, A. 1995. On not being anywhere near the "project": revolutionary ways of putting ourselves in the picture, *Antipode*, 27, 137–66.

Christopherson, S. 1989: On being outside "the project." *Antipode*, 21, 83–9.

Clifford, J. 1997: Spatial practices: fieldwork, travel, and the discipline of anthropology. In A. Gupta and J. Ferguson (eds), *Anthropological Locations: Boundaries and Grounds of a Field Science*. Berkeley: University of California Press, 185–219.

Cloke, P., Philo, C. and Sadler, D. 1991: *Approaching Human Geography: An Introduction to Contemporary Theoretical Debates*. London: Paul Chapman.

Cochrane, A. 1998: Illusions of power: interviewing local elites. *Environment and Planning A*, 30, 2121–32.

Collay, M. 1989: How does researcher questioning technique influence participant response in qualitative research? Paper presented at the annual meeting of the American Educational Research Association, San Francisco, March.

Collins, P. H. 1998: *Fighting Words: Black Women and the Search for Justice*. Durham, N. C.: Duke University Press.

Connell, R. 1987: *Gender and Power*. Cambridge: Polity Press.

Connell, R. 1995: *Masculinities*. Berkeley: University of California Press.

Cope, M. 1998: "She hath done what she could": community, citizenship, and place among women in late nineteenth century Colorado. *Historical Geography*, 26, 45–64.

Davies, C. 1999: *Reflexive Ethnography: A Guide to Researching Selves and Others*. London and New York: Routledge.

Davis, K. (ed.), 1997: *Embodied Practices. Feminist Perspectives on the Body.* Thousand Oaks, Calif.: Sage.

Dawes, K. 1999: *Natural Mysticism: Towards a New Reggae Aesthetic in Caribbean Writing.* Leeds: Peepal Tree Press.

Demeritt, D. 2000: The new social contract for science: accountability, relevance and value in US and UK science and research policy. *Antipode*, 32, 308–29.

Denzin, N. 1978: *The Research Act.* New York: McGraw-Hill.

Denzin, N. K. and Lincoln, Y. S. (eds) 1994: *Handbook on Qualitative Research.* Thousand Oaks, Calif.: Sage.

Denzin, N. K. and Lincoln, Y. S. (eds) 1998a: *Collecting and Interpreting Qualitative Materials.* Thousand Oaks, Calif.: Sage.

Denzin, N. K. and Lincoln, Y. S. (eds) 1998b: *Strategies of Qualitative Inquiry.* Thousand Oaks, Calif.: Sage.

Denzin, N. K. and Lincoln, Y. S. (eds) 1998c: *The Landscape of Qualitative Research: Theories and Issues.* Thousand Oaks, Calif.: Sage.

Desbiens, C. 1999: Feminism "in" geography: elsewhere, beyond and the politics of paradoxical space. *Gender, Place and Culture*, 6, 179–85.

Devault, M. L. 1990: Talking and listening from women's standpoint: feminist strategies for interviewing and analysis. *Social Problems*, 37, 96–116.

Devault, M. L. 1999: *Liberating Method: Feminism and Social Research.* Philadelphia: Temple University Press.

di Leonardo, M. (ed.) 1991: *Gender at the Crossroads of Knowledge.* Berkeley: University of California Press.

Domosh, M. 1996: *Invented Cities: The Creation of Landscape in Nineteenth-Century New York and Boston.* New Haven: Yale University Press.

Domosh, M. 1997: Geography and gender: the personal and the political. *Progress in Human Geography*, 21, 81–7.

Du Plessis, R. and Fougere, G. 1995: The social world of "The Piano." *Sites*, 31, Spring, 132–43.

Dyck, I. 1993: Ethnography: a feminist method. *Canadian Geographer*, 37, 52–7.

Dyck, I. and McLaren, A. T. 2001: Telling it like it is . . . or just another story? Tales of immigrant settlement. Unpublished manuscript. Available from Isabel Dyck, School of Rehabitation Sciences, University of British Columbia, Vancouver, BC, Canada.

Edwards, R. 1990: Connecting method and epistemology: a white woman interviewing black women. *Women's Studies International Forum*, 13, 477–90.

Eichler, M. 1988: *Nonsexist Research Methods: A Practical Guide.* Boston: Allen and Unwin.

Eichler, M. 1991: *Nonsexist Research Methods: A Practical Guide.* Revised edn. Boston: Allen and Unwin.

Ellis, C. 1997: Evocative autoethnography: writing emotionally about our lives. In W. Tierney and Y. Lincoln (eds), *Representation and the Text: Re-framing the Narrative Voice*, Albany, NY: State University of New York, 115–39.

El-Or, T. 1992: Do you really know how they make love? The limits on intimacy with ethnographic informants. *Qualitative Sociology*, 15, 53–71.

Employment Equity Act – http://info.load-otea.hrdc-drhc.gc.ca/~weeweb/lege.htm

England, K. 1993: Suburban pink collar ghettos: the spatial entrapment of women? *Annals, Association of American Geographers*, 83, 25–42.

England, K. 1994: Getting personal: reflexivity, positionality and feminist research. *The Professional Geographer*, 46, 80–9.

England, K. 1995: "Girls in the office": Job search and recruiting in a local clerical labor market, *Environment and Planning A*, 27: 1995–2018.

Evans, M. 1988: Participant observation. The researcher as research tool. In J. Eyles and D. M. Smith (eds), *Qualitative Methods in Human Geography*. Cambridge, UK: Polity Press, 197–218.

Ewick, P. 1994: Integrating feminist epistemologies in undergraduate research methods. *Gender and Society*, 8, 92–108.

Faludi, S. 1992: *Backlash: The Undeclared War Against Women*. London: Vintage.

Finch, J. 1984: "It's great to have someone to talk to": the ethics and politics of interviewing women. In C. Bell and H. Roberts (eds), *Social Researching: Politics, Problems, Practice*, London: Routledge Kegan Paul, 70–87.

Fonow, M. M. and Cook, J. A. 1991: *Beyond Methodology: Feminist Scholarship as Lived Research*. Bloomington: Indiana University Press.

Foucault, M. 1980: *Power/Knowledge: Selected Interviews and Writings 1972–1977*. New York: Pantheon.

Fox, B. J. 1988: Conceptualizing patriarchy. *Canadian Review of Sociology and Anthropology*, 25, 163–82.

Friedman, S. S. 1985: Authority in the feminist classroom: a contradiction in terms? In M. Culley and C. Portuges (eds) *Gendered Subjects*, Boston: Routledge and Kegan Paul, 203–08.

Frohlick, S. 1999: "Home has always been hard for me": single mothers' narratives of identity, home and loss. In R. Bridgman, S. Cole and H. Howard-Bobiwash (eds), *Feminist Fields: Ethnographic Insights*. Peterborough, Ont.: Broadview Press, 86–102.

Gahan, C. and Hannibal, M. 1998: *Doing Qualitative Research Using QSR NUD*IST*. Thousand Oaks, Calif.: Sage.

Gebhardt, E. 1978: A critique of methodology. In A. Arato and E. Gebhardt (eds), *The Essential Frankfurt School Reader*. Oxford: Blackwell, 371–406.

Geographie heute 1982: Frauen und Entwicklung [Women and Development], Vol. 14.

Gibson, K. 1996: Social polarisation and the politics of difference: discourses in collision or collusion. In K. Gibson et al.(eds), *Restructuring Difference: Social Polarisation and the City*. Melbourne: Australian Housing Urban Research Institute, Working Paper No. 6, 5–12.

Gibson-Graham, J.-K. 1994: "Stuffed if I know!" Reflections on post-modern feminist social research. *Gender, Place and Culture*, 1, 205–24.

Gibson-Graham, J.-K. 1996: *The End of Capitalism (As We Knew It)*. Oxford: Blackwell.

Gilbert, A-F. 1985: Frauenforschung am Beispiel der Time-Geography. [Women's Research: An Example of Time-Geography]. Magistra thesis, University of Zürich, Department of Geography.

Gilbert, A-F. and Rössler, M. 1982: Quer durch die Geographie in halsbrecherischen Sprüngen auf den Spuren der Frauen. [Tremendous Jumps in the Search of Women throughout Geography] *Geoscop*, 37, 4–13.

Gilbert, M. 1994: The politics of location: doing feminist research at "home." *Professional Geographer*, 46, 90–6.

Gilbert, M. and Masucci, M. 1999: Information Management and Technology Use Workshop for Economic and Environmental Community Organizations. Workshop Report, Philadelphia, PA, March 25–27. Available from the authors at Department of Geography, Temple University, Philadelphia, PA, USA.

Gilmartin, M. 1999: The Irish travels of Asenath Nicholson in 1844–45. In A. Buttimer, S. D. Brunn and U. Wardenga (eds) *Text and Image: Social Construction of Regional Knowledges*. Leipzig: Selbstverlag Institut für Länderkunde, 248–55.

Glacken, C. 1967: *Traces on the Rhodian Shore: Nature and Culture in Western Thought from Ancient Times to the End of the Eighteenth Century*. Berkeley and Los Angeles: University of California Press.

Glaser, B. 1978: *Advances in the Methodology of Grounded Theory. Theoretical Sensitivity*. San Francisco: The Sociology Press.

Glaser, B. and Strauss, A. 1967: *The Discovery of Grounded Theory*. Chicago: Aldine.

Gluck, S. B. and Patai, D. (eds) 1991: *Women's Words: The Feminist Practice of Oral History*. London and New York: Routledge.

Gordon, P., Kumar, A. and Richardson, H. W. 1989: Gender differences in metropolitan travel behaviour. *Regional Studies*, 23, 499–510.

Gottfried, H. (ed.) 1996: *Feminism and Social Change: Bridging Theory and Practice*. Urbana and Chicago: University of Illinois Press.

Green, J. and Hart, L. 1999: The impact of context on data. In R. Barbour and J. Kitzinger(eds), *Developing Focus Group Research: Politics, Theory and Practice*. Thousand Oaks, Calif.: Sage, 21–35.

Grosz, E. 1994: *Volatile Bodies: Towards a Corporeal Feminism*. Bloomington and Indianapolis: Indiana University Press.

Grosz, E. and Probyn, E. (eds) 1995: *Sexy Bodies. The Strange Carnalities of Feminism*. London and New York: Routledge.

Guerrero, S. H. 1999: *Gender-sensitive and Feminist Methodologies: A Handbook for Health and Social Researchers*. Dillman, Quezon City: University Center for Women's Studies, University of the Philippines.

Gupta, A. and Ferguson, J. (eds) 1997: *Anthropological Locations: Boundaries and Grounds of a Field Science*. Berkeley and Los Angeles: University of California Press.

Halford, S., Savage, M. and Witz, A. 1997: *Gender, Careers and Organisations*. London: Macmillan.

Hall, S. 1991: Old and new identities, old and new ethnicities. In A. D. King (ed.), *Culture, Globalization, and the World System*. Hampshire: Macmillan Press, 41–68.

Hanson, S. 1997: As the world turns: new horizons in feminist geographic methodologies. In J. P. Jones III, H. J. Nast and S. M. Roberts (eds), *Thresholds in Feminist Geography: Difference, Methodology, Representation*. Boulder, CO: Rowman and Littlefield, 119–28.

Hanson, S. 2000: Networking. *Professional Geographer*, 52, 751–8.

Hanson, S. and Pratt, G. 1990: Geographic perspectives on the occupational segregation of women. *National Geographic Research*, 6, 376–99.

Hanson, S. and Pratt, G. 1995: *Gender, Work and Space*. London and New York: Routledge.

Haraway, D. 1988: Situated knowledges: the science question in feminism and the privilege of partial perspective, *Feminist Studies*, 14, 575–99.

Haraway, D. J. 1991: *Simians, Cyborgs, and Women: The Reinvention of Nature*. London and New York: Routledge.

Harding, S. 1986: *The Science Question in Feminism*. Ithaca: Cornell University Press.

Harding, S. (ed.) 1987a: *Feminist Methodology*. Bloomington: Indiana University Press.

Harding, S. 1987b: Introduction: is there a feminist method? In S. Harding (ed.), Feminist Methodology. Bloomington: Indiana University Press, 1–14.

Harding, S. 1991: *Whose Science? Whose Knowledge?: Thinking from Women's Lives*. Ithaca: Cornell University Press.

Harding, S. 1992: Rethinking standpoint epistemology: what is "strong objectivity"? *Centennial Review*, 36, 437–70.

Harding, S. 1998: Can men be subjects of feminist thought? In T. Digby (ed.), *Men Doing Feminism*, London and New York: Routledge, 171–96.

Hartsock, N. 1984: *Money, Sex and Power*. Boston: Northeastern University Press.

Harvey, L. 1990: *Critical Social Research*. London: Unwin Hyman.

Hawkesworth, M. 1989: Knowers, knowing, known: feminist theory and claims of truth. *Signs*, 14, 533–57.

Hawkesworth, M. E. 1990: *Beyond Oppression: Feminist Theory and Political Strategy*. New York: Continuum.

Hekman, S. J. 1999: *The Future of Differences: Truth and Method in Feminist Theory*. Malden, Mass.: Polity Press.

Herod, A. 1999: Reflections on interviewing foreign elites: praxis, positionality, validity, and the cult of the insider. *Geoforum*, 30, 313–27.

Hertz, R. 1997: Introduction: reflexivity and voice. In R. Hertz (ed), *Reflexivity and Voice*. Thousand Oaks, Calif.: Sage, vii–xviii.

Hertz R. and Imber, B. (eds) 1995: *Studying Elites Using Qualitative Methods*. Thousand Oaks, Calif.: Sage.

Hesse-Biber, S. J., Gilmartin, C. K. and Lydenberg, R. (eds) 1999: *Feminist Approaches to Theory and Methodology: An Interdisciplinary Reader*. New York: Oxford University Press.

Hiebert, D. Creese, G. Dyck, I., Ley, D,. McLaren, A., Pratt, G. 1998: Immigrant Settlement in Greater Vancouver: An Introduction to the Community Studies. Working Paper on the Metropolis webpage: www.riim.metropolis.net

Historical Geography. 1998: Gender and the city, 26, 1–91.

Holloway, S. and Valentine, G. 2000: *Children's Geographies: Playing, Living, Learning*. London: Routledge.

hooks, b. 1984: *Feminist Theory: From Margin to Centre*. Boston: South End Press.

hooks, b. 1990: *Yearning: Race, Gender, and Cultural Politics*. London: Turnaround.

hooks, b. 1992: *Black Looks: Race and Representation*. Toronto: Between the Lines.

Hopkins, M. C. 1993: Is anonymity possible? Writing about refugees in the United States. In C. B. Brettell (ed.), *When They Read What We Write: The Politics of Ethnography*. Westport, Conn. and London: Bergin and Garvey, 121–9.

HRDC (Human Resources Development Canada) 1998: *Annual Report: Employment Equity Act, 1998*. Hull: HRDC.

Hudson, B. 1977: The new geography and the new imperialism: 1870–1918. *Antipode*, 9(2), 12–19.

Hughes, A. 1999: Constructing economic geographies from corporate interviews: insights from a cross-country comparison of retailer-supplier relationships. *Geoforum*, 30, 363–74.

Hurren, W. 1998: Living with/in the lines: poetic possibilities for world writing. *Gender, Place and Culture*, 5, 301–4.

Jackson, P. 1985: Urban ethnography. *Progress in Human Geography*, 9, 157–76.

Jackson, P. 1991: The cultural politics of masculinity: towards a social geography. *Transactions of the Institute of British Geographers*, 16, 199–213.

Jackson, P. 1993: Changing ourselves: A geography of position. In R. J. Johnston (ed.), *The Challenge for Geography*. Oxford: Blackwell, 198–214.

Jaggar, A. M. 1983: *Feminist Politics and Human Nature*. Totowa, N.J.: Rowman and Allanheld.

Jarosz, L. 1999: A feminist political ecology perspective. *Gender, Place and Culture*, 6, 390–3.

Jayaratne, T. E. 1983: The value of quantitative methodology for feminist research. In G. Bowles and R. D. Klein (eds), *Theories of Women's Studies*. London: Routledge and Kegan Paul, 140–61.

Jayaratne, T. E. and Stewart, A. J. 1991: Quantitative and qualitative methods in the social sciences: current feminist issues and practical strategies. In M. M. Fonow and J. A. Cook (eds), *Beyond Methodology: Feminist Scholarship as Lived Research*. Bloomington, IN: Indiana University Press, 85–106.

Johnson, H. 2000: Of discourse, dialogue and dutiful daughters, *Resources for Feminist Research/Documentation sur la recherche féministe*, 28 (1–2), 229–43.

Johnson, L. C. 1989: Feminist or gender geography in Australasia? *Journal of Geography in Higher Education*, 13, 85–8.

Johnson, L. C. 1990: New patriarchal economies in the Australian textile industry. *Antipode*, 22, 1–32.

Johnson, L. C. 1993: Text-ured brick: Speculations on the cultural production of domestic space, *Australian Geographical Studies*, 31, 201–13.

Johnson, L. C. 1994a: Occupying the suburban frontier: accommodating difference on Melbourne's urban fringe. In A. Blunt and G. Rose (eds), *Writing Women and Space: Colonial and Postcolonial Geographies*. New York: Guildford Press, 141–68.

Johnson, L. C. 1994b: Colonising the suburban frontier: place-making on Melbourne's urban fringe. In S. Watson and K. Gibson (eds), *Metropolis Now:*

Planning and the Urban in Contemporary Australia. Sydney: Pluto Press, 46–59.

Johnson, L. C. 1996: Restructuring and Socio-Economic Polarisation in a Regional Industrial Centre. In K. Gibson, et al. (eds) *Restructuring Difference: Social Polarisation and the City.* Melbourne: Australian Housing Urban Research Institute, Working Paper No. 6, 43–57.

Johnson, L. C. 1997: Regions matter! In understanding employment change. In *Taskforce 2000 2nd Report.* Melbourne, Australian Fabian Society Pamphlet No. 54, 21–8.

Johnson, L. C. and Wright, S. 1994: (White) papering over the regional problem: unemployment in Geelong and the federal government response. *Australian Geographer,* 25(2), 121–5.

Johnson, L. C. with Jacobs, J. and Huggins, J. 2000: *Placebound: Australian Feminist Geographies.* South Melbourne: Oxford University Press.

Johnson, R. J., Gregory, D., Pratt, G. and Watts, M. (eds) 2000: *The Dictionary of Human Geography.* (4th edn). Oxford: Blackwell.

Johnson-Hill, J. A. 1995: *I-Sight: The World of Rastafari: An Interpretive Sociological Account of Rastafarian Ethics.* Lanham, Md. and London: Scarecrow Press.

Johnston-Anumonwo, I. 1995: Racial differences in commuting behavior of women in Buffalo, 1980–1990. *Urban Geography,* 16, 23–45.

Johnston-Anumonwo, I. 1997: Race, gender, and constrained work trips in Buffalo, NY, 1990. *Professional Geographer,* 49, 306–17.

Jones, J. P. III, Nast, H. J. and Roberts, S. M. (eds) 1997a: Part 1: Difference, in *Thresholds in Feminist Geography.* Latham, MD: Rowman and Littlefield, 1–115.

Jones, J. P. III, Nast, H. J. and Roberts, S. M. (eds) 1997b: *Thresholds in Feminist Geography.* Latham, MD: Rowman and Littlefield.

Kahane, D. J. 1998: Male feminism as oxymoron? In T. Digby (ed.), *Men Doing Feminism.* London and New York: Routledge, 213–36.

Katz, C. 1994: Playing the field: questions of fieldwork in geography. *Professional Geographer,* 46, 67–72.

Katz, C. 1996: The expeditions of conjurers: ethnography, power, and pretense. In D. Wolf (ed.), *Feminist Dilemmas in Fieldwork.* Boulder, Colo: Westview Press, 170–84.

Katz, C. and Monk, J. (eds) 1993: *Full Circles: Geographies of Women over the Life Course.* London and New York: Routledge.

Kawabata, H. 1997: Gender, ethnicity and space: the case of racial ethnic minority women telecommuters in Omaha, Nebraska. Unpublished M.A. thesis. Department of Geography, University of Nebraska, Omaha.

Kilduff M. and Mehra, A. 1997: Postmodernism and organizational research. *Academy of Management Review,* 22, 453–81.

Kimmel, M. S. 1998: Who's afraid of men doing feminism? In T. Digby (ed.), *Men Doing Feminism.* London and New York: Routledge, 57–68.

Kirby, S. and McKenna, K. 1989: *Experience research social change: Methods from the Margins.* Toronto: Garamond Press.

Kirsch, G. R. 1999: *Ethical Dilemmas in Feminist Research: The Politics of*

Location, Interpretation, and Publication. Albany: State University of New York Press.

Kitzinger, J. 1994: The methodology of focus groups: the importance of interaction between research participants. *Sociology of Health and Illness,* 16, 103–21.

Kobayashi, A. 1994: Coloring the field: gender, "race" and the politics of fieldwork. *The Professional Geographer,* 46, 73–80.

Kobayashi, A. 1997: The paradox of difference and diversity (or, why the threshold keeps moving). In J. P. Jones III, H. J. Nast and S. M. Roberts (eds), *Thresholds in Feminist Geography: Difference, Methodology, Representation.* Lanham, MD: Rowman and Littlefield, 3–9.

Kobayashi, A. and Peake, L. 1994: Unnatural discourse: "race" and gender in geography. *Gender, Place and Culture,* 1, 225–43.

Kofman, E. 1998: Whose city? Gender, class, and immigration in globalizing European cities. In R. Fincher and J. M. Jacobs (eds), *Cities of Difference.* London and New York: Guilford, 279–300.

Krueger, R. A. 1994: *Focus Groups: A Practical Guide for Applied Research.* Thousand Oaks, Calif.: Sage.

Kruks, S. 2000: *Retrieving Experience: Subjectivity and Recognition in Feminist Politics.* Ithaca: Cornell University Press.

Kvale, S. 1996: *Interviews: An introduction to qualitative research interviewing.* Thousand Oaks, Calif.: Sage.

Kwan, M.-P. 1998: Space-time and integral measures of individual accessibility: a comparative analysis using a point-based framework. *Geographical Analysis,* 30 (3), 191–216.

Kwan, M.-P. 1999a: Gender, the home-work link, and space-time patterns of nonemployment activities. *Economic Geography,* 75, 370–94.

Kwan, M.-P. 1999b: Gender and individual access to urban opportunities: a study using space-time measures. *Professional Geographer,* 51, 210–27.

Kwan, M.-P. 2000a: Other GISs in other worlds: feminist visualization and re-envisioning GIS. Paper presented at the Annual Meeting of the Association of American Geographers, Pittsburgh, PA, April.

Kwan, M.-P. 2000b: Gender differences in space-time constraints. *Area,* 32, 145–56.

Laclau, E. and Mouffe, C. 1985: *Hegemony and Socialist Strategy: Towards a Radical Democratic Politics.* London: Verso.

Lather, P. 1986: Research as praxis. *Harvard Educational Review,* 56, 257–76.

Lather, P. 1991: *Getting Smart. Feminist Research and Pedagogy With/In the Postmodern.* London and New York: Routledge.

Laurie, N. 1999: "More than the blood of earth mothers." *Gender, Place and Culture,* 6, 393–400

Laws, S. 1986: The social meaning of menstruation: a feminist investigation. Unpublished Ph.D. dissertation. Warwick University, UK. Cited in H. Gottfried 1996: Engaging Women's Communities: Dilemmas and contradictions in feminist research. In H. Gottfried (ed.) *Feminism and Social Change.* Urbana and Chicago: University of Chicago Press, p. 5.

Lawson, V. 1995: The politics of difference: examining the quantitative/qualitative dualism in post-structuralist feminist research. *Professional Geographer*, 47, 449–57.

Little, J. and Austin P. 1996: Women and the rural idyll. *Journal of Rural Studies*, 12, 101–11.

Livingstone, D. 1992: *The Geographical Tradition*. Oxford and Cambridge, MA: Blackwell.

Lloyd, G. 1984: *The Man of Reason: Male and Female in Western Philosophy*. London: Methuen.

Longhurst, R. 1995: The body and geography. *Gender, Place and Culture*, 2, 97–105.

Lorde, A. 1984: *Sister Outsider*. Trumansburg, NY: The Crossing Press.

Luff, D. 1999: Dialogue across the divides: "moments of rapport" and power in feminist research with anti-feminist women. *Sociology*, 33, 687–703.

Luzzadder-Beach, S. and MacFarlane, A. M. 2000: The environment of gender and science: status and perspectives of women and men in physical geography. *The Professional Geographer*, 52, 407–24.

Lyons, L. and Chipperfield, J. 2000: (De)constructing the interview: a critique of the participatory model, *Resources for Feminist Research/Documentation sur la recherche féministe*, 28 (1–2), 33–48.

Madge, C. and Bee, A. 1999: Women, science and identity: interviews with female physical geographers. *Area*, 31, 335–48.

Madge, C., Raghuram, P., Skelton, T., Willis, K. and Williams, J. 1997: Methods and methodologies in feminist geographies: politics, practice and power. In Women and Geography Study Group, *Feminist Geographies. Explorations in Diversity and Difference*. Essex: Addison Wesley Longman, 86–111.

Magris, C. 1999: *Danube*. London: Collins Harvill.

Maguire, P. 1987: *Doing Participatory Research: A Feminist Approach*. Amherst, Mass.: The Center for International Education, University of Massachusetts.

Marcus, G. E. 1998: *Ethnography Through Thick and Thin*. Princeton: Princeton University Press.

Marshall, J. 1999: Insiders and outsiders: the role of insularity, migration and modernity on Grand Manan, New Brunswick. In R. King and J. Connell (eds), *Small Worlds, Global Lives: Islands and Migration*. London and New York: Pinter, 95–113.

Massey, D. 1993: Power geometry and a progressive sense of place. In J. Bird, B. Curtis, T. Putman, G. Robertson and L. Tickner (eds), *Mapping the Future: Local Cultures, Global Change*. London: Routledge, 59–69.

Mathison, S. 1988: Why triangulate? *Education Researcher*, 17, 13–17.

Mattingly, D. and Falconer Al-Hindi, K. 1995: Should women count? A context for the debate. *Professional Geographer*, 47, 427–35.

McDowell, L. 1992a: Doing gender: feminism, feminists and research methods in human geography. *Transactions, Institute of British Geographers*, 17, 399–416.

McDowell, L. 1992b: Multiple voices: speaking from inside and outside "the project." *Antipode*, 24, 56–72.

McDowell, L. 1992c: Valid games? A response to Erica Schoenberger. *Professional Geographer*, 44, 212–15.

McDowell, L. 1993a: Space, place and gender relations: part 1. Feminist empiricism and the geography of social relations. *Progress in Human Geography*, 17, 157–79.

McDowell, L. 1993b. Space, place and gender relations: part 2. Identity, difference, feminist geometries and geographies. *Progress in Human Geography*, 17, 305–18.

McDowell, L. 1997a: *Capital Culture: Gender at Work in the City*. Oxford: Blackwell.

McDowell, L. 1997b: Women/gender/feminists: doing feminist geography. *Journal of Geography in Higher Education*, 21, 381–400.

McDowell, L. 1998: Elites in the city of London: some methodological considerations. *Environment and Planning A*, 30, 2133–46.

McDowell, L. 1999: *Gender, Identity and Place: Understanding Feminist Geographies*. Cambridge: Polity Press.

McDowell, L. and Bowlby, S. 1983: Teaching feminist geography. *Journal of Geography in Higher Education*, 7, 97–107.

McDowell, L. and Sharp, J. P. 1997: *Space, Gender, Knowledge: Feminist Readings*. London: Arnold.

McDowell, L. and Sharp, J. P. 1999: *A Feminist Glossary of Human Geography*. London: Edward Arnold.

McLafferty, S. 1995: Counting for women. *Professional Geographer*, 47, 436–42.

McLafferty, S. and Preston, V. 1992: Spatial mismatch and labor market segmentation for African American and Latino women. *Economic Geography*, 68, 406–31.

McLafferty, S. and Preston, V. 1996: Spatial mismatch and employment in a decade of restructuring. *Professional Geographer*, 48, 420–31.

McLafferty, S. and Preston, V. 1997: Gender, race, and the determinants of commuting: New York in 1990. *Urban Geography*, 18 (3), 192–212.

Michell, L. 1999: Combining focus groups and interviews: telling how it is; telling how it feels. In R. Barbour and J. Kitzinger (eds), *Developing Focus Group Research: Politics, Theory and Practice*, Thousand Oaks, Calif.: Sage, 36–46.

Middleton, S. 1985: Feminism and Education in Post-War New Zealand: A Sociological Analysis. Unpublished PhD dissertation. University of Waikato, New Zealand.

Miles, M. and Huberman, A. 1994: *Qualitative Data Analysis. An Expanded Sourcebook*. Thousand Oaks, Calif.: Sage.

Mohanty, C. T. 1991: Under western eyes: feminist scholarship and colonial discourses. In C. T. Mohanty, A. Russo and L. Torres (eds), *Third World Women and the Politics of Feminism*. Bloomington: Indiana University Press, 51–80.

Monk, J. 1985: Feminist transformation: how can it be accomplished? *Journal of Geography in Higher Education*, 9, 101–5.

Monk, J. 1997: Marginal notes on representations. In J. P. Jones, III, H. J. Nast and S. M. Roberts (eds), *Thresholds in Feminist Geography*. Lanham, Md.: Rowman and Littlefield, 241–53.

Monk, J. and Hanson, S. 1982: On not excluding half of the human in human geography. *The Professional Geographer*, 34, 11–23.

Moraga, C. and Anzaldúa, G. (eds) 1981. *This Bridge Called My Back: Writings by Radical Women of Color*. Watertown, Mass.: Persephone.

Morgan, D. L. 1997: *Focus Groups as Qualitative Research* (2nd edn). Thousand Oaks, Calif.: Sage.

Morgan, D. L. 1998: *The Focus Group Guidebook: Focus Group Kit 1*. Thousand Oaks, Calif.: Sage.

Morgan, K. P. 1992: The perils and paradoxes of feminist pedagogy. In D. Shogan (ed.), *A Reader in Feminist Ethics*. Toronto: Canadian Scholars' Press, 393–405.

Morrison, T. 1992: *Playing in the Dark: Whiteness and the Literary Imagination*. Boston and London: Harvard University Press.

Morrison, T. 1997: *Paradise*. New York: Alfred A. Knopf.

Moss, P. 1993: Introductory comments, *Canadian Geographer*, 37, 48–9.

Moss, P. 1995a: Embeddedness in practice, numbers in context: the politics of knowing and doing. *Professional Geographer*, 47, 442–9.

Moss, P. 1995b: Reflections on the "gap" as part of the politics of research design. *Antipode*, 27, 82–90.

Moss, P. 1999: Autobiographical notes on chronic illness. In E. K. Teather (ed.), *Mind and Body Spaces*, London: Routledge.

Moss, P. and Matwychuk, M. 1996: Rebordering feminist praxis. *Atlantis*, 21 (1), 3–9.

Moss, P. and Matwychuk, M. 2000: Beyond speaking as an "as a" and stating the "etc." Toward a praxis of difference. *Frontiers*, 21 (3), 82–104.

Moss, P. and McMahon, M. 2000: Between a flake and a strident bitch: making "it" count in the academy, *Resources for Feminist Research/Documentation sur la recherche féministe*, 28 (1–2), 15–32.

Moss, P., De Bres, K., Cravey, A., Hyndman, J., Hirschboek, K. and Masucci, M. 1999: Mentoring as feminist praxis: strategies for ourselves and others. *Journal of Geography in Higher Education*, 23, 413–27.

Mullings, B. 1999: Insider or outsider, both or neither: some dilemmas of interviewing in a cross-cultural setting *Geoforum*, 30, 337–50.

Myers, G. and Macnaghten, P. 1999: Can focus groups be analysed as talk? In R. Barbour and J. Kitzinger (eds), *Developing Focus Group Research: Politics, Theory and Practice*, Thousand Oaks, Calif.: Sage, 173–85.

Myerson, G. and Rydin, Y. 1996: *The Language of Environment. A New Rhetoric*. London: UCL Press.

Nadel-Klein, J. 1997: Crossing a representational divide: from west to east in Scottish ethnography. In A. James, J. Hockey and A. Dawson (eds), *After Writing Culture: Epistemology and Praxis in Contemporary Anthropology*. London and New York: Routledge, 86–102.

Nagar, R. 1997: Exploring methodological borderlands through oral narratives. In J. P. Jones III, H. J. Nast and S. M. Roberts (eds), *Thresholds in Feminist Geography: Difference, Methodology, Representation*. Lanham, MD: Rowman and Littlefield, 203–24.

Nairn, K. 1998: Disciplining Identities: Gender, Geography and the Culture of Fieldtrips. Unpublished PhD dissertation. University of Waikato, New Zealand.

Nairn, K. 1999: Embodied fieldwork. *Journal of Geography*, 98, 272–82.

Napia, E. B., Ram, R., Ward, K., Wright, D. H. and Heperi, V. L. 1999: Ka hao te rangatahi (a new? net goes a fishing): Our new voices are really old as we play around with academic things like writing and research. Presentation at Reclaiming Voice II: Ethnographic Inquiry and Qualitative Research in a Postmodern Age, (June), Irvine, CA.

Naples, N. A. (ed) 1998: *Community Activism and Feminist Politics: Organizing Across Race, Class, and Gender*. London and New York: Routledge.

Nast, H. J. 1994: Women in the field: critical feminist methodologies and theoretical perspectives. *The Professional Geographer*, 46, 54–66.

Nast, H. J. 1998: The body as "place": reflexivity and fieldwork in Kano, Nigeria. In H. J. Nast and S. Pile (eds), *Places Through the Body*. London and New York: Routledge, 93–116.

Newton, E. 1993: My best informant's dress: The erotic equation in fieldwork. *Cultural Anthropology*, 8(1), 3–23.

Ní Dhomhnaill, N. 1996: Dinnsheanachas: the naming of high or holy places. In P. Yaeger (ed.), *The Geography of Identity*. Ann Arbor: University of Michigan Press, 408–32.

Nicotera, A. 1999: The woman academic as subject/object/self: dismantling the illusion of duality. *Communication Theory*, 9, 430–64.

Nielson, J. M. 1990: *Feminist Research Methods: Exemplary Readings in the Social Sciences*. Boulder, Colo.: Westview Press.

Norwood, V. and Monk, J. (eds) 1987: *The Desert is No Lady: Southwestern Landscapes in Women's Writing and Art*. New Haven: Yale University Press.

Oakley, A. 1981: Interviewing women: a contradiction in terms. In H. Roberts (ed.), *Doing Feminist Research*. New York: Routledge and Kegan Paul, 30–61.

Oakley, A. 2000: *Experiments in Knowing: Gender and Method in the Social Sciences*. New York: New Press.

Oberhauser, A. 1997: The home as "field": households and homework in rural Appalachia. In J. P. Jones III, H. J. Nast and S. M. Roberts (eds), *Thresholds in Feminist Geography: Difference, Methodology, Representation*. Latham, MD: Rowman and Littlefield, 165–82.

Okely, J. 1992: Anthropology and autobiography: participatory experience and embodied knowledge. In J. Okely and H. Callaway (eds), *Anthropology and Autobiography*. London and New York: Routledge, 1–28.

Okely, J. 1996: *Own or Other Culture*. London: Routledge.

Okoko, E. 1999: Women and environmental change in the Niger Delta, Nigeria: evidence from Ibeno. *Gender, Place and Culture*, 6, 373–8.

Opie, A. 1992: Qualitative research: appropriation of the "other" and empowerment. *Feminist Review*, 40, 52–69.

Ortner, S. B. 1995: Resistance and the problem of ethnographic refusal. *Society for Comparative Study of Society and History*, 37, 173–92.

Ostheider, M. 1984: Geographische Frauenforschung – Ein neuer theoretischer Ansatz? [Geographical Women Research – A New Theoretical Issue?]. In G. Bahrenberg and W. Taubmann (eds), *Bremer Beiträge zur Geographie und Raumplanung, Vol. 5*. Breman: University of Bremen, 202–26. Reprinted in: S.

Bock, U. Hünlein, H. Klamp, and M. Treske (eds) 1989: *Frauen(t)räume* [*Women's spaces/dreams*], *Urb et regio* Vol. 52. Kassel: Kasseler Schriften zu Geographie und Planung, 10–44.

Parry, B. 1998: Hunting the gene-hunters: the role of hybrid networks, status, and chance in conceptualising and accessing "corporate elites." *Environment and Planning A*, 30, 2147–62.

Pateman, C. 1989: *The Disorder of Women: Democracy, Feminism and Political Theory.* Cambridge: Polity Press.

Peace, R. and Longhurst, R. 1997: Producing feminist geography "down under." *Gender Place and Culture*, 4, 115–19.

Peake, L. 1983: Teaching feminist geography: another perspective. *Journal of Geography in Higher Education*, 7, 186–9.

Perrons, D. 1999: Missing subjects? Searching for gender in official statistics. In D. Dorling and S. Simpson (eds), *Statistics in Society: The Arithmetic of Politics.* London: Arnold, 105–14.

Personal Narratives Group. 1989: *Interpreting Women's Lives: Feminist Theory and Personal Narratives.* Bloomington: Indiana University Press.

Phoenix, A. 1994: Practising feminist research: the intersection of gender and race in the research process. In M. Maynard and J. Purvis (eds), *Researching Women's Lives From a Feminist Perspective.* London and New York: Routledge.

Poland, B. D. 2001: Transcript quality. In J. Gubrium and J. Holstein (eds), *Handbook of Interviewing.* Thousand Oaks, Calif.: Sage.

Popper, K. 1959: *The Logic of Scientific Discovery.* London: Hutchinson.

Porter, M. 1995: Call yourself a sociologist – and you've never even been arrested? *Canadian Review of Sociology and Anthropology*, 32(4), 415–38.

Pratt, G. 1989: Quantitative techniques and humanistic-historical materialist perspectives. In A. Kobayashi and S. Mackenzie (eds), *Remaking Human Geography.* Boston: Unwin Hyman, 101–15.

Pratt, G. 1993: Reflections on poststructuralism and feminist empirics, theory and practice, *Antipode*, 25, 51–63.

Pratt, G. 1997: Stereotypes and ambivalence: nanny agent's stereotypes of domestic workers in Vancouver, B.C. *Gender, Place and Culture*, 4, 159–77.

Pratt, G. 1999: From Registered Nurse to Registered Nanny: Discursive Geographies of Filipina Domestic Workers in Vancouver, B.C. *Economic Geography*, 75, 215–36.

Pratt, G. 2000: Research Performance. *Environment and Planning D: Society and Space*, 18, 639–51.

Pratt, G. in collaboration with the Philippine Women Centre. 1998: Inscribing domestic work on Filipina bodies. In H. Nast and S. Pile (eds), *Places Through the Body*, London: Routledge, 283–304.

Pratt, G. in collaboration with the Philippine Women Centre. 1999: "Is this Canada?" Domestic workers' experiences in Vancouver, B.C. In J. Momsen (ed.), *Gender, Migration and Domestic Service*, New York: Routledge, 23–42.

Pratt, M. L. 1992: *Imperial Eyes: Travel Writing and Transculturation.* London and New York: Routledge.

Pratt, M. L. 1994: Transculturation and autoethnography: Peru 1615/1980. In F.

Barker, P. Hulme and M. Iverson (eds), *Colonial Discourse/Postcolonial Theory*, Manchester: Manchester University Press, 24–46.

Pred, A. 1984: Place as historically contingent process: Structurationism and the time-geography of becoming places. *Annals of the Association of American Geographers* 74, 279–97.

Price, P. L. 1999: Bodies, faith, and inner landscapes: rethinking change from the very local. *Latin American Perspectives* 26 (3), 37–59.

Professional Geographer. 1994: Women in the field. 46, 54–102.

Professional Geographer. 1995: Should women count? The role of quantitative methodology in feminist geographic research. 47, 426–66.

Professional Geographer. 2000: Women in geography in the 21st Century. 54, 697–738.

Pronger, B. 1998: On your knees: carnal knowledge, masculine dissolution, doing feminism. In T. Digby (ed.), *Men Doing Feminism*. London and New York: Routledge, 69–80.

Psathas, G. 1995: *Conversation Analysis: The Study of in-Talk Analysis*. Thousand Oaks, Calif.: Sage.

Psathas, G. and Anderson, T. 1990: The "practices" of transcription in conversation analysis. *Semiotica*, 78 (1–2), 75–99.

Pugh, A. 1990: My statistics and feminism – a true story. In L. Stanley (ed), *Feminist Praxis: Research, Theory and Epistemology in Feminist Sociology*. London: Routledge, 103–12.

Raghuram, P., Madge, C. and Skelton, T. 1998: Feminist research methodologies and student projects in geography. *Journal of Geography in Higher Education*, 22, 35–48.

Reay, D. 1996: Dealing with difficult differences: reflexivity and social class in feminist research. *Feminism and Psychology*, 6, 443–56.

Reed-Danahay, D. 1999: Introduction. In D. Reed-Danahay (ed.), *Auto/Ethnography: Rewriting the Self and the Social*. New York: Berg, 1–11.

Reichert, D. 1994: Woman as utopia: against relations of representation. *Gender, Place and Culture*, 1, 91–102.

Reid, C. 2000: Seduction and enlightenment in feminist action research, *Resources for Feminist Research/Documentation sur la recherche féministe*, 28 (1–2), 169–88.

Reinharz, S. 1992: *Feminist Methods in Social Research*. Oxford: Oxford University Press.

Resources for Feminist Research/Documentation sur la recherche féministe. 2000: Feminist Qualitative Research/Féminisme et recherche qualitative. 28 (1–2), 9–243.

Ribbens, J. 1989: Interviewing women – an unnatural situation? *Women's Studies International Forum*, 12, 579–92.

Ristock, J. L. and Pennell, J. 1997: *Community Research as Empowerment: Feminist Links, Postmodern Interruptions*. Toronto: Oxford University Press.

Roberts, H. (ed.) 1981: *Doing Feminist Research*. New York: Routledge and Kegan Paul.

Robinson, J. 1994: White women researching/representing "others": from antiapartheid to postcolonialism? In A. Blunt and G. Rose (eds), *Writing Women and Space*. New York and London: Guilford Press, 197–225.

Robson, E. 1999: Problematising oil and gender in Nigeria. *Gender, Place and Culture*, 6, 379–90.

Rose, D. 1993: On feminism, method and methods in human geography: an idiosyncratic overview. *Canadian Geographer*, 37, 57–61.

Rose, G. 1992: Geography as a science of observation: the landscape, the gaze and masculinity. In F. Driver and G. Rose (eds), *Nature and Science: Essays in the History of Geographical Knowledge*. Historical Geography Research Series, 28, February. Cheltenham: Historical Geography Research Group.

Rose, G. 1993: *Feminism and Geography: The Limits of Geographical Knowledge*. Minneapolis: University of Minnesota Press.

Rose, G. 1997: Situating knowledges: postionality, reflexivities and other tactics. *Progress in Human Geography*, 21, 305–20.

Rosenbloom, S. and Burns, E. 1994: Why working women drive alone: implications for travel reduction programs. *Transportation Research Record*, 1459, 39–45.

Said, E. 1978: *Orientalism*. New York: Vintage.

Samarasinghe, V. 1997: Counting women's work: the intersection of time and space. In J. P. Jones III, H. J. Nast and S. M. Roberts (eds), *Thresholds in Feminist Geography: Difference, Methodology, Representation*. New York: Rowman and Littlefield, 129–44.

Sandoval, C. 2000: *Methodology of the Oppressed*. Minneapolis: University of Minnesota Press.

Sayer, A. 1985: *Method in Social Science: A Realist Approach*. London: Hutchinson.

Sayer, A. and Morgan, K. 1985: A modern industry in a declining region: links between method, theory and policy. In D. Massey and R. Meegan (eds), *Politics and Method: Contrasting Studies in Industrial Geography*. London: Methuen, 147–68.

Schoenberger, E. 1991: The corporate interview as a research method in economic geography. *Professional Geographer*, 43, 180–9.

Schoenberger, E. 1992: Self-criticism and self-awareness in research: a reply to Linda McDowell. *Professional Geographer*, 44, 215–18.

Schoenberger, E. 1994: Corporate strategy and corporate strategists: power identity, and knowledge within the firm. *Environment and Planning A*, 26, 435–51.

Scott, J. C. 1990: *Domination and the Arts of Resistance: Hidden Transcripts*. New Haven: Yale University Press.

Seager, J. and Olson, S. 1986: *Women in the World: An International Atlas*. London: Pluto.

Sedgwick, E. 1990: *Epistemology of the Closet*. Berkeley: University of California Press.

Seidler, V. J. 1991: *Recreating Sexual Politics: Men, Feminism and Politics*. London and New York: Routledge.

Sheppard, E. 2001: Quantitative geography: representations, practices and possibilities. *Environment and Planning D: Society and Space*, 18, forthcoming.

Skelton, T. and Valentine, G. 1998: *Cool Places: Geographies of Cool Cultures*. London: Routledge.

Smith, D. 1990a: *Texts, Facts, and Femininity: Exploring the Relations of Ruling*. London and New York: Routledge.

Smith, D. 1990b: *The Conceptual Practices of Power: A Feminist Sociology of Knowledge*. Boston: Northeastern University Press.

Soja, E. 1980: The socio-spatial dialectic. *Annals of the Association of American Geographers*, 70, 207–25.

Song, M. and Parker, D. 1995: Commonality, difference and the dynamics of disclosure in in-depth interviewing. *Sociology*, 29, 241–56.

Sparke, M. 1994: Writing on patriarchal missiles: the chauvinism of the gulf war and the limits of critique. *Environment and Planning A*, 26, 1061–89.

Sparke, M. 1996: Displacing the field in fieldwork: masculinity, metaphor and space. In N. Duncan (ed.), *BodySpace: Destabilizing Geographies of Gender and Sexuality*. London and New York: Routledge, 212–33.

Spender, D. 1981: *Men's Studies Modified*. Oxford: Pergamon Press.

Spivak, G. C. 1988: *In Other Worlds: Essays in Cultural Politics*. London: Routledge.

Spivak, G. C. 1990: Questions of multi-culturalism. In S. Harasym (ed.), *The Postcolonial Critic: Interviews, Strategies, Dialogues*. London and New York: Routledge, 59–66.

Sprague, J. and Zimmerman, M. K. 1989: Quality and quantity: reconstructing feminist methodology. *The American Sociologist*, 20, 71–86.

Stacey, J. 1988: Can there be a feminist ethnography? *Women's Studies International Forum*, 11, 21–7.

Stacey, J. 1990: *Brave New Families*. New York: Basic Books.

Stacey, J. 1996: Can there be a feminist ethnography? In H. Gottfried (ed.), *Feminism and Social Change*, Urbana and Chicago: University of Illinois Press, 88–104.

Stanley, L. 1992: *The Auto/Biographical I/Eye*. Manchester: Manchester University Press.

Stanley, L. and Wise, S. 1983: *Breaking Out: Feminist Ontology and Epistemology*. London, Routledge.

Stanley, L. and Wise, S. 1993: *Breaking Out Again: Feminist Ontology and Epistemology*. London and New York: Routledge.

Stoddart, D. R. 1986: *On Geography*. Oxford: Blackwell.

Swann, C. 1997: Reading the bleeding body: discourses of premenstrual syndrome. In J. Ussher (ed.), *Body Talk: The Material and Discursive Regulation of Sexuality, Madness and Reproduction*. London and New York: Routledge, 176–98.

Tekülve, M. 1985: Zur Lebenssituation von Frauen in der Dritten Welt am Beispiel des ländlichen Indiens. Eine Fallstudie aus zwei Dörfern des Thanjavur-Distrikts. [Women's Lives in the Third World: An Example from Rural India. A Case Study of Two Villages in the Thanjavur District]. Magistra thesis, University of Göttingen, Department of Geography.

Tesch, R. 1991: Introduction to Computers and Qualitative Data II Special Issue, Part I. *Qualitative Sociology*, 14 (3), 225–43.

Thiele, L. 1990: The agony of politics: the Nietzschean roots of Foucault's thought, *American Political Science Review*, 84, 907–25.

Thomas, R. J. 1995: Interviewing important people in big companies. In R. Hertz and B. Imber (eds), *Studying Elites Using Qualitative Methods*. Thousand Oaks, Calif.: Sage, 3–17.

Tivers, J. 1985: *Women Attached: The Daily Lives of Women with Young Children*. London: Croom Helm.

Tuan, Y.-F. 1976: Humanistic geography. *Annals, Association of American Geographers*, 66, 266–76.

Tuana, N. (ed.) 1993: *Feminism and Science*. Bloomington: Indiana University Press.

Twyman, C., Morrison, J. and Sporton, D. 1999: The final fifth: autobiography, reflexivity and interpretation in cross-cultural research. *Area*, 31, 313–25.

Useem, M. 1995: Reaching corporate executives. In R. Hertz R. and B. Imber (eds), *Studying Elites Using Qualitative Methods*. Thousand Oaks, Calif.: Sage, 18–39.

Ussher, J. 1997: *Fantasies of Femininity: Reframing the Boundaries of Sex*. London and New York: Routledge.

Valentine, G. 1993: (Hetero)sexing space: lesbian experiences of everyday space. *Environment and Planning D: Society and Space*, 11, 395–413.

Valentine, G. 2000: *Consuming Geographies: We Are Where We Eat*. London: Routledge.

Valentine, G. 2001: *Social Geographies: Space and Society*. Harlow: Prentice Hall.

Visweswaran, K. 1994: *Fictions of Feminist Ethnography*. Minneapolis: University of Minnesota Press.

Wagner, I. 1985: Frauen in den Naturwissenschaften: Institutionelle und kognitive Widerstände [Women in the natural sciences: institutional and experiential resistance]. In P. Feierabend and C. Thomas (eds), *Grenzprobleme der Wissenschaften*, Zürich: Verlag der Fachvereine, 215–25.

Waldby, C. 1995: Destruction: boundary erotics and refigurations of the heterosexual male body. In E. Grosz and E. Probyn (eds), *Sexy Bodies. The Strange Carnalities of Feminism*. London: Routledge, 266–77.

Wallman, S. 1978: The boundaries of "race": Processes of ethnicity in England. *Man*, 13, 200–17.

Ward, K. G. and Jones, M. 1999: Researching local elites: reflexivity, "situatedness" and political-temporal contingency. *Geoforum*, 30, 301–12.

Wasserfall, R. 1993: Reflexivity, feminism and difference. *Qualitative Sociology*, 16, 23–41.

Wasserfall, R. 1997: Reflexivity, feminism, and difference. In R. Hertz (ed), *Reflexivity and Voice*. Thousand Oaks, Calif.: Sage, 150–68.

WGSG (Women and Geography Study Group) 1984: *Geography and Gender*. London: Hutchinson.

WGSG (Women and Geography Study Group) 1997: *Feminist Geographies: Explorations in Diversity and Difference*. Harlow: Longman.

Wilkinson, S. 1999: How useful are focus groups in feminist research? In R. Barbour and J. Kitzinger (eds), *Developing Focus Group Research: Politics, Theory and Practice*, London: Sage, 64–78.

Williams, C. L. and Heikes, E. J. 1993: The importance of researcher's gender in the in-depth interview. *Gender and Society*, 7, 280–91.

Wolf, D. (ed.) 1996: *Feminist Dilemmas in Fieldwork*. Boulder, Colo.: Westview Press.

Woods, M. 1998: Rethinking elites: networks, space, and local politics. *Environment and Planning A*, 30, 2101–19.

Young, I. 1990: The ideal of community and the politics of difference. In L. Nicholson (ed.), *Feminism/Postmodernism*. London and New York: Routledge, 300–23.

Index